NUREG-1930, Supplement 1

# Safety Evaluation Report

Related to the License Renewal of Indian Point Nuclear Generating Unit Nos. 2 and 3

## Supplement 1

Docket Nos. 50-247 and 50-286

**Entergy Nuclear Operations, Inc.**

**United States Nuclear Regulatory Commission**

Office of Nuclear Reactor Regulation

August 2011

THIS PAGE INTENTIONALLY LEFT BLANK.

# ABSTRACT

This document is a supplemental safety evaluation report (SSER) for the license renewal application for Indian Point Nuclear Generating Unit Nos. 2 and 3 (IP2 and IP3) as filed by Entergy Nuclear Operations, Inc., (Entergy or the applicant). By letter dated April 23, 2007, and as supplemented by letters dated May 3 and June 21, 2007, the applicant submitted the LRA in accordance with Title 10, Part 54, of the *Code of Federal Regulations*, "Requirements for Renewal of Operating Licenses for Nuclear Power Plants." Entergy requests renewal of the IP2 and IP3 operating licenses (Facility Operating License Numbers DPR-26 and DPR-64, respectively) for a period of 20 years beyond the current expirations at midnight on September 28, 2013, for IP2, and at midnight on December 12, 2015, for IP3.

The NRC staff published NUREG-1930 "Safety Evaluation Report Related to the License Renewal of Indian Point Nuclear Generating Unit Nos. 2 and 3" in two volumes in November 2009, which summarized the results of its safety review of the renewal application for compliance with the requirements of Title 10, Part 54, of the *Code of Federal Regulations,* (10 CFR Part 54), "Requirements for Renewal of Operating Licenses for Nuclear Power Plants." This SSER documents the staff's review of supplemental information provided by the applicant since the issuance of the SER. This information includes annual updates required by 10 CFR 54.21(b), and updated information and commitments in response to staff requests for additional information. This document only discusses the changes to the SER.

**THIS PAGE INTENTIONALLY LEFT BLANK**

# TABLE OF CONTENTS

# Appendices

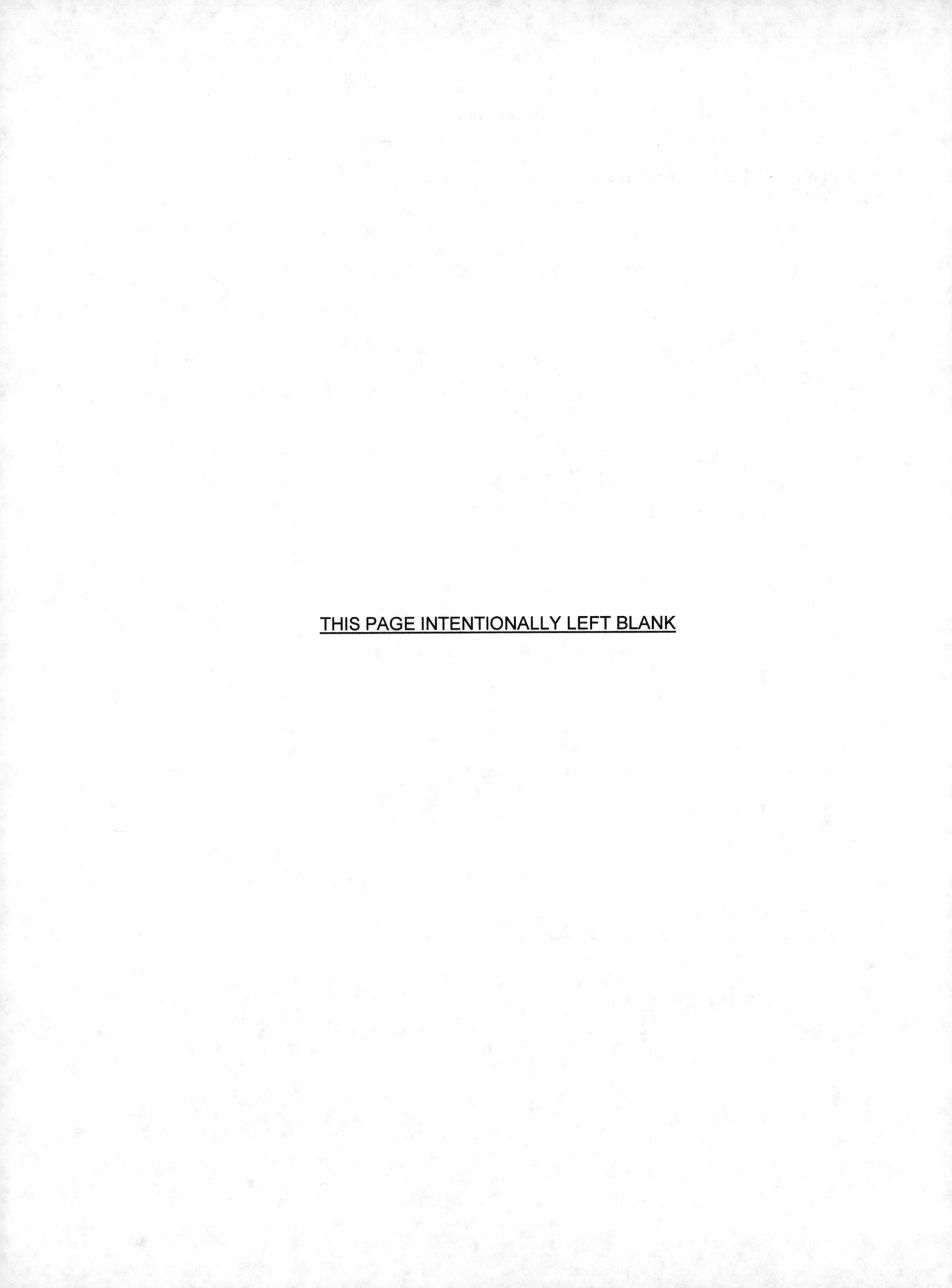

THIS PAGE INTENTIONALLY LEFT BLANK

# ABBREVIATIONS

ACI                 American Concrete Institute
ADAMS          Agencywide Documents Access and Management System
AMP              aging management program
AMR              aging management review
ASME            American Society of Mechanical Engineers
AWWA           American Water Works Association

CAP              corrective action program
CASS            cast austenitic stainless steel
CLB              current licensing basis
CFR              Code of Federal Regulations
CUF              cumulative usage factor

EPRI             Electric Power Research Institute
EQ                environmental qualification

$F_{en}$                environmental fatigue life correction factor

GALL            Generic Aging Lessons Learned
GL                Generic Letter

IPEC             Indian Point Energy Center
IP2 and IP3    Indian Point Nuclear Generating Unit Nos. 2 and 3
ISI                inservice inspection

kV                kilovolts

LER              licensee event report
LRA              license renewal application
LWR             light water reactor

MRP             Materials Reliability Program

NRC             Nuclear Regulatory Commission

OEP             operating experience program

PEO             period of extended operation
PVC             polyvinyl chloride
PWSCC        primary water stress corrosion cracking

QA               Quality Assurance

# ABBREVIATIONS

| | |
|---|---|
| RAI | request for additional information |
| RCPB | reactor coolant pressure boundary |
| RIS | Regulatory Information Summary |
| | |
| SCC | stress corrosion cracking |
| SCs | Structures and Components |
| SER | safety evaluation report |
| SG | steam generator |
| SGMP | steam generator management program |
| SRP-LR | Standard Review Plan for License Renewal Applications for Nuclear Power Plants |
| SSCs | systems, structures and components |
| SSER | supplemental safety evaluation report |
| | |
| UFSAR | updated final safety analysis report |
| USAS | United States of America Standard |
| | |
| V | volts |

# SECTION 1

# INTRODUCTION AND GENERAL DISCUSSION

## 1.1 Introduction

This document is a supplemental safety evaluation report (SSER) for the license renewal application (LRA) for Indian Point Nuclear Generating Unit Nos. 2 and 3 (IP2 and IP3) as filed by Entergy Nuclear Operations, Inc., (Entergy or the applicant). By letter dated April 23, 2007, and as supplemented by letters dated May 3 and June 21, 2007, the applicant submitted the LRA in accordance with Title 10, Part 54, of the *Code of Federal Regulations*, "Requirements for Renewal of Operating Licenses for Nuclear Power Plants." Entergy requests renewal of the IP2 and IP3 operating licenses (Facility Operating License Numbers DPR-26 and DPR-64, respectively) for a period of 20 years beyond the current expirations at midnight on September 28, 2013, for IP2, and at midnight on December 12, 2015, for IP3.

The NRC staff published NUREG-1930 "Safety Evaluation Report Related to the License Renewal of Indian Point Nuclear Generating Unit Nos. 2 and 3" in two volumes in November 2009, which summarized the results of its safety review of the renewal application for compliance with the requirements of Title 10, Part 54, of the *Code of Federal Regulations,* (10 CFR Part 54), "Requirements for Renewal of Operating Licenses for Nuclear Power Plants." This SSER documents the staff's review of supplemental information provided by the applicant since the issuance of the SER. This information includes annual updates required by 10 CFR 54.21(b), and updated information and commitments in response to staff requests for additional information (RAIs). This SSER supplements portions of SER Sections 3, 4, 5, Appendix A, and Appendix B.

# SECTION 2

## STRUCTURES SYSTEMS AND COMPONENTS SUBJECT TO AGING MANAGEMENT REVIEW

There are no changes or updates to this section of the safety evaluation report.

THIS PAGE INTENTIONALLY LEFT BLANK

# SECTION 3

# AGING MANAGEMENT REVIEW RESULTS

## 3.0  Applicant's Use of the Generic Aging Lessons Learned Report

### 3.0.3  Aging Management Programs

### 3.0.3.1  AMPs Consistent with the GALL Report

3.0.3.1.2 Buried Piping and Tanks Inspection Program

Summary of Technical Information in the Application.  By letters dated March 28, July 14, and July 21, 2011, the applicant supplemented the LRA to include revisions to the Buried Piping and Tanks Inspection Program.  The applicant revised the number of planned inspections of buried piping within the scope of license renewal from 45 (a non-specific mix of direct and indirect inspections) in the 10-year period prior to the period of extended operation to 31 direct inspections of steel piping, 3 direct inspections of stainless steel piping, and 17 indirect inspections in the 10-year period prior to the period of extended operation.  The applicant also revised the number of inspections to be conducted in the period of extended operation from a non-specific total in the first 10 years of the period of extended operation based on the results of inspections to be completed prior to the period of extended operation to 28 direct inspections of steel piping and two direct inspections of stainless steel piping in each 10-year period of the period of extended operation.  The applicant stated that until such time that an alternative inspection methodology is qualified and demonstrated to be effective, it will not use alternate inspection methods to inspect buried piping within the scope of license renewal for the stated minimum number of inspections.  The staff noted that based on this response, the applicant will not use the guided wave method in lieu of excavated direct visual inspection of buried piping within the scope of license renewal.

Staff Evaluation.  In light of operating experience that has occurred coincident with and after the staff's initial evaluation of the Indian Point Nuclear Generating Unit Nos. 2 and 3 (IP2 and IP3) license renewal application (LRA) and issuance of the SER, the staff reviewed the LRA and SER and determined that additional information was required to evaluate how the applicant is implementing changes to its program based on more recent industry experience.  By letter dated February 10, 2011 the staff issued request for additional information (RAI) 3.0.3.1.2-1 requesting that the applicant:

 a) state how many buried piping segments are within the scope of license renewal for each material; how many are code and/or safety-related; and, how many with the potential to release materials detrimental to the environment, will be inspected

 b) confirm which systems have cathodic protection, whether periodic NACE surveys are conducted, and the availability of the cathodic protection system

 c) state why piping that is not cathodically protected will meet the minimum design wall thickness throughout the period of extended operation

d) provide details on buried pipe inspections that have been conducted since July 2009 that provide insights on the quality of backfill in the vicinity of buried pipes

e) state what alternative inspection methods will be used in lieu of direct visual inspection of buried pipe

f) state what systems have underground piping (i.e., piping and tanks that are below grade, contained in a vault or tunnel, in contact with air, and where access for inspection is restricted)

g) state what soil parameters will be included in soil corrosivity testing, how often, and in what locations sampling will be conducted; and how localized soil conditions will be factored into the number of inspections of buried piping with the scope of license renewal.

In its response dated March 28, 2011, the applicant stated that:

a) The following excavated direct visual inspections will be conducted on buried piping within the scope of license renewal:

  - at IP2, at least nine locations of code or safety-related steel piping and 11 locations of steel piping containing hazardous materials will be inspected in the 10-year period prior to the period of extended operation,

  - at IP3, at least eight locations of code or safety-related steel piping, three locations of steel piping containing hazardous materials, and three locations of stainless steel piping containing hazardous materials will be inspected in the 10-year period prior to the period of extended operation,

  - in summary, prior to the period of extended operation, there will be 31 inspections of steel piping and three inspections of stainless steel piping

  - at IP2, at least six locations of code or safety-related steel piping and eight locations of steel piping containing hazardous materials will be inspected each 10-year period during the period of extended operation,

  - at IP3, at least six locations of code or safety-related steel piping, eight locations of steel piping containing hazardous materials, and two locations of stainless steel piping containing hazardous materials will be inspected each 10-year period during the period of extended operation,

  - In summary, during each 10-year period during the period of extended operation, there will be 28 inspections of steel piping and two inspections of stainless steel piping,

  - with the exception of ten inspections that have already been conducted, each excavated direct visual examination inspection will consist of a minimum of 10 linear feet of pipe for the full circumference. The 10 inspections conducted to date ranged from five feet to more than ten feet with an average of eight linear feet,

- the non safety-related security generator buried propane tank and steel piping will be age managed by monitoring the level of the propane in the tank; however, opportunistic inspections of the tank will be performed.

b) The buried city water piping in the vicinity of the Algonquin gas pipelines is the only buried piping within the scope of license renewal that is cathodically protected. Annual NACE surveys are conducted. Cathodic protection availability since installation in November 2009 has exceeded 98 percent.

c) The applicant stated that there is reasonable assurance that buried piping within the scope of license renewal will continue to meet its design function without cathodic protection because: (1) recent inspections have found the piping's coating to be in good condition, (2) soil resistivity measurements have shown the soil to be non-aggressive, (3) risk ranking of inspection locations has been and will be used to identify those areas most susceptible to corrosion, (4) further soil samples will be obtained with the number of inspections being increased if the soil is corrosive, and (5) an adequate number of inspections have been conducted to date and are planned.

d) Two 10-foot sections of 16-inch pipe and an eight foot section of 10-inch city water piping were inspected in October 2009. Eight feet of fire protection piping was inspected in November 2009. The backfill did not contain rocks or foreign material that would damage external coatings. The coatings were found to be in good condition.

e) It has no plans to use volumetric inspection methods in lieu of direct visual examination of buried pipe. In the absence of a qualified method and until such time that one is demonstrated to be effective, it will not use alternate inspection methods.

f) It has no underground piping or tanks (i.e., piping and tanks below grade, contained in a vault or tunnel, in contact with air, where access for inspection is restricted) within the scope of license renewal.

g) Both soil resistivity and American Water Works Association (AWWA) C105 rating factors will be used to determine soil corrosivity. Soil parameters utilized in the AWWA C105 rating are resistivity, pH, redox potential, sulfides, and moisture. If soil resistivity is less than 20,000 ohm-cm or the AWWA C105 rating exceeds 10, the soil will be considered corrosive. Soil samples will be taken prior to the period of extended operation and at least once every 10 years in the period of extended operation. Soil samples will be taken at a minimum of two locations near buried piping within the scope of license renewal to obtain representative soil conditions for each system. The number of inspections of steel code or safety-related piping will be increased from six to eight inspections and for steel piping containing hazardous materials from eight to 12 if the soil is corrosive.

The staff finds the applicant's response to RAI 3.0.3.1.2-1, items (b), (c), (d), (e), (f), and (g) acceptable. Although the service water, containment isolation support, auxiliary feedwater, plant drains, fuel oil, security diesel propane, and fire protection systems are not cathodically protected, the applicant's response is acceptable in that:

- The applicant is risk informing its piping inspection locations to select those with the greatest potential for leakage.

- The applicant is sampling the soil for corrosivity prior to and during the period of extended operation, using standard industry methodologies to determine soil corrosivity, and will be increasing the number of inspections if the soil is corrosive.

- Steel piping is coated.

- Recent inspections found that the backfill did not contain rocks or foreign material that would damage external coatings and the coatings were found to be in good condition. The staff noted that foreign material in backfill caused sufficient damage of the condensate storage tank return line coating such that the line corroded and leaked, and in other instances inspections found coating damage; however, the applicant's proposed number of inspections meet the current staff position for number of inspections for a plant with no cathodic protection and unacceptable backfill quality.

The staff finds that the applicant is conducting a sufficient number of inspections to establish a reasonable basis for the staff to conclude that the current licensing basis (CLB) function(s) of the buried systems within the scope of license renewal will be maintained throughout the period of extended operation. The staff noted that the applicant will not use alternative inspection methodologies unless they are qualified and demonstrated to be effective. The staff also noted that the applicant has no underground piping or tanks (i.e., piping and tanks below grade, contained in a vault or tunnel, in contact with air, where access for inspection is restricted) within the scope of license renewal.

The staff's concern described in item (a) of RAI 3.0.3.1.2-1 was not resolved because the applicant proposed to manage the effects of aging for the buried propane piping and tanks by monitoring tank level. The license renewal function of these components is to provide a pressure boundary function for the tanks and piping to support a 10 CFR Part 50 Appendix R event. By letter dated June 15, 2011, the staff issued RAI 3.0.3.1.2-2 requesting that the applicant state the basis for why monitoring propane tank level ensures that the license renewal function of the tank and piping is met.

In its response dated July 14, 2011, the applicant stated that: there are two propane tanks and 50 feet of piping that supply fuel to one engine that supports fire protection and 10 CFR Part 50 Appendix R lighting requirements; the tanks and piping are not safety-related and do not contain hazardous materials; the piping and tanks are constructed of the same material, coated, and exposed to similar soil conditions as the other buried steel piping within the scope of license renewal which will have multiple direct visual inspections as described above; observation of tank levels allows personnel to initiate corrective actions to repair or replace a leaking tank; and monitoring of propane tank level is similar to the recommendation in NUREG-1801, "Generic Aging Lessons Learned (GALL) Report," Revision 2 which allows monitoring of jockey pump operation to age manage buried fire system piping within the scope of license renewal.

The staff finds the applicant's response acceptable because the applicant's 31 direct visual inspections of buried steel piping prior to the period of extended operation and 28 during each of the 10-year periods in the period of extended operation, coupled with the buried propane tanks and piping being constructed of similar material and exposed to a similar environment, will establish objective evidence of the condition of the buried propane piping and tanks. The staff's concern described in RAI 3.0.3.1.2-2 is resolved.

Based on its review of the applicant's response to RAI 3.0.3.1.2-1 and RAI 3.0.3.1.2-2, the staff finds that operating experience related to the applicant's program demonstrates that it can adequately manage the detrimental effects of aging on systems, structures and components (SSCs) within the scope of the program, and implementation of the program will result in the applicant taking appropriate corrective actions. The staff confirmed that the "operating experience" program element satisfies the criteria in NUREG-1800, Revision 1, "Standard Review Plan for Review of License Renewal Applications for Nuclear Power Plants" (SRP-LR) Section A.1.2.3.10 and, therefore, the staff finds it acceptable.

Operating Experience. There are no changes or updates to this section of the SER.

Updated Final Safety Analysis Report (UFSAR) Supplement. The staff noted that the applicant had not updated UFSAR Sections A.2.1.5 and A.3.1.5, as required by 10 CFR 54.21(d), to reflect the response to RAI 3.0.3.1.2-1 in regard to the number and frequency of piping inspections and soil testing. Therefore, by letter dated June 15, 2011, the staff issued RAI 3.0.3.1.2-3 requesting that the applicant revise the UFSAR supplement to reflect the number and frequency of inspections and soil testing planned for all buried pipe within the scope of license renewal.

In its response dated July 14, 2011, and amended by letter dated July 27, 2011, the applicant revised LRA Sections A.2.1.5 and A.3.1.5 to reflect the number and frequency of piping inspections and soil testing.

The staff finds the applicant's response acceptable because the UFSAR supplement establishes the number and frequency of piping inspections and soil testing licensing basis for the program. The staff's concern described in RAI 3.0.3.1.2-3 is resolved.

Conclusion On the basis of its review of the applicant's response to RAIs 3.0.3.1.2-1, 3.0.3.1.2-2, and 3.0.3.1.2-3, the staff finds that those program elements for which the applicant claimed consistency with the GALL Report are consistent. The staff concludes that the applicant has demonstrated that the effects of aging will be adequately managed so that the intended function(s) will be maintained consistent with the CLB for the period of extended operation, as required by 10 CFR 54.21(a)(3). The staff also reviewed the UFSAR supplement for this AMP and concludes that it provides an adequate summary description of the program, as required by 10 CFR 54.21(d).

3.0.3.1.6 Non-EQ Inaccessible Medium-Voltage Cable Program

Summary of Technical Information in the Application. LRA Section B.1.23 as modified by letters dated March 28, 2011 and July 7, 2011, describes the Non-EQ (environmental qualification) Inaccessible Medium-Voltage Cable Program as a new program that will be consistent with the

GALL Report aging management program (AMP) XI.E3, "Inaccessible Medium-Voltage Cables Not Subject to 10 CFR 50.49 Environmental Qualification Requirements."

The applicant stated that the Non-EQ Inaccessible Medium-Voltage Cable Program includes periodic inspections for water collection in cable manholes, and provides for the testing of cables. In-scope medium-voltage cables (i.e., cables with operating voltage from 400 volts (V) to 35 kilovolts (kV)) exposed to significant moisture will be tested at least every 6 years for an indication of the condition of the conductor insulation. The program inspects for water accumulation in manholes at least annually.

Staff Evaluation. The staff's evaluation of the applicant's Non-EQ Inaccessible Medium-Voltage Cable Program is documented in SER Section 3.0.3.1.6. Subsequent to issuance of the SER, the staff issued RAI 3.0.3.1.6-1 requesting that the applicant respond to the following to address recent industry operating experience with inaccessible low and medium voltage power cables:

1. Explain how Entergy Nuclear Operations, Inc., (Entergy) will manage the effects of aging on inaccessible low voltage power cables within the scope of license renewal and subject to aging management review, including consideration of recently identified industry operating experience and any plant-specific operating experience. The discussion should include assessment of the aging management program description, program elements (i.e., scope of program, parameters monitored/inspected, detection of aging effects, and corrective actions), and an UFSAR summary description to demonstrate reasonable assurance that the intended functions of inaccessible low voltage power cables subject to adverse localized environments will be maintained consistent with the current licensing basis through the period of extended operation.

2. Provide an evaluation showing that the proposed Non-EQ Inaccessible Medium-Voltage Cable Program test and inspection frequencies, including event-driven inspections, incorporate recent industry and plant-specific operating experience for both inaccessible low and medium voltage cables.

3. In Commitment 40, Entergy committed to evaluate plant-specific and industry operating experience prior to entering the period of extended operation. Explain how the proposed Non-EQ Inaccessible Medium-Voltage Cable Program will continue to ensure that future industry and plant-specific operating experience will be incorporated into the program such that inspection and test frequencies may be increased based on test and inspection results.

By letters dated March 28, 2011, and July 7, 2011, the applicant provided a supplement to the license renewal application including the management of inaccessible low voltage power cables. The applicant stated that it will include low-voltage power cables (400V to 2kV) in the Non-EQ Inaccessible Medium-Voltage Cable Program. In addition, the applicant stated that the Non-EQ Inaccessible Medium-Voltage Cable Program will include provisions to increase cable testing and manhole inspections frequencies, including adjusting frequencies based on the results of testing and inspections.

Specifically, the applicant stated that the Non-EQ Inaccessible Medium-Voltage Cable Program will be based on and consistent with GALL AMP. XI.E3, "Inaccessible Medium-Voltage Cables

Not Subject to 10 CFR 50.49 Environmental Qualification Requirements" with the following enhancements described in LRA Sections B.1.23, A.2.1.22 and A.3.1.22:

- The significant voltage exposure definition (applicable to medium voltage cable (2kV to 35kV) subjected to system voltage for more than 25 percent of the time) is removed as a scope of program criterion.

- The Non-EQ Inaccessible Medium-Voltage Cable Program is expanded to include 400V to 2kV in-scope inaccessible low voltage power cable.

- The performance of manhole inspections is increased to at least annually.

- The testing of inaccessible cables (400V to 35kV) for degradation of cable insulation will be performed at least once every six years.

- Event driven inspections (e.g. heavy rain or flooding), are incorporated into the Non-EQ Inaccessible Medium-Voltage Cable Program.

- Cable test results include reviews to determine the need for more frequent testing.

- Manhole inspection results are evaluated to determine the need for more frequent manhole inspections.

The applicant also provided more recent plant operating experience stating that the applicant's response to Generic Letter (GL) 2007-01 "Inaccessible or Underground Power Cable Failures that Disable Accident Mitigation Systems or Cause Plant Transients" described that IP3 had experienced two cable failures, and IP2 had experienced no failures based on the scoping criteria set forth in GL 2007-01. The applicant explained that both IP3 failures involved low-voltage power cables but the failures were due to mechanical damage and not aging. The applicant also stated that a review of plant operating experience since the applicant's response to GL 2007-01 identified one IP2 failure and no IP3 failures of low or medium voltage power cables within scope of the maintenance rule or license renewal. The applicant further stated that the cable failure identified for IP2 was caused during modification activities and was not aging related.

With the information provided by the applicant's RAI response, the staff finds the applicant's Non-EQ Inaccessible Medium-Voltage Cable Program acceptable with respect to inaccessible low voltage power cable because the applicant has included inaccessible low voltage power cables (400v to 2kV) within the scope of this program consistent with industry and plant-specific operating experience such that there is reasonable assurance that inaccessible low voltage power cables subject to significant moisture will be adequately managed during the period of extended operation. The applicant also revised cable testing frequencies to once every 6 years and manhole inspections to at least annually with added event-driven inspections for in-scope manholes. The applicant's incorporation of increased testing and inspection frequencies and event driven inspections into the Non-EQ Inaccessible Medium-Voltage Cable Program is acceptable because the changes are also consistent with industry operating experience. The elimination of the significant voltage criterion definition (subjected to system voltage for more than 25 percent of the time) is also acceptable because this change expands the scope of the program consistent with inaccessible medium voltage cable operating experience. The applicant also revised the Non-EQ Inaccessible Medium-Voltage Cable Program to provide more frequent inspection and test frequencies as necessary based on inspection and test results, and current staff recommendations. The staff's concerns related to inaccessible power cables operating experience described in RAI 3.0.3.1.6-1 are resolved.

Operating Experience. There are no changes or updates to this section of the SER.

UFSAR Supplement. In LRA Sections A.2.1.22 and A.3.1.22, the applicant provided the UFSAR supplement for the Non-EQ Inaccessible Medium-Voltage Cable Program. By response to RAI 3.0.3.1.6-1 dated March 28, 2011 and supplemented by letters dated July 14, and July 27, 2011 the applicant revised the UFSAR supplements A.2.1.22 and A.3.1.22 to include inaccessible power cable (400V to 2kV) and to address operating experience for inaccessible power cable within the scope of license renewal. The March 28, 2011 letter included a revision to the UFSAR supplement for IP2 (LRA Section A.2.1.22) which stated that "In addition to the periodic manhole inspections, inspection of event-driven occurrences, such as heavy rain or flooding will be performed." The July 14, 2011 letter contained an UFSAR supplement for IP3 (LRA Section A.3.1.22) which stated that "The inspection frequency for water collection is established and performed based on plant specific operating experience with cable wetting or submergence in manholes (i.e. the inspection is performed periodically based on water accumulation over time and event driven occurrences, such as heavy rain or flooding). The July 27, 2011 letter revised the IP2 UFSAR supplement to correspond with the IP3 UFSAR supplement provided in the July 14, 2011 letter. It was unclear to the staff if the revised UFSAR supplements included event driven inspections. Therefore, the staff held a teleconference with the applicant on August 3, 2011, to request clarification on the revised UFSAR supplements. The applicant indicated that the Non-EQ Inaccessible Medium-Voltage Cable Program, as described in the revised UFSAR supplements, would not include event driven inspections. The staff indicated on the teleconference that the program should be able to ensure that the cables were not submerged in the intervals between periodic inspections. Therefore, by letter dated August 9, 2011 the applicant revised the UFSAR supplements for IP2 and IP3 to state that event driven inspections will be performed. The staff reviewed these sections, including the applicant's RAI response, and determines that the information in the UFSAR supplement is an adequate summary description of the program, as required by 10 CFR 54.21 (d). The applicant committed to implement the Non-EQ Inaccessible Medium-Voltage Cable Program prior to the period of extended operation. The applicant further stated that this new program will be implemented consistent with LRA Section B.1.23 and the corresponding program described in GALL AMP XI.E3, "Inaccessible Medium-Voltage Cables Not Subject to 10 CFR 50.49 Environmental Qualification Requirements" (Commitment 15).

By letter dated July 27, 2009, the applicant added a new commitment (Commitment 40) which states that plant-specific and appropriate industry operating experience will be evaluated and lessons learned will be used to establish appropriate monitoring and inspection frequencies to assess aging effects for the new aging management programs prior to the period of extended operation. In RAI 3.0.3.1.6-1 the staff asked the applicant to explain how the proposed Non-EQ Inaccessible Medium-Voltage Cable Program will continue to ensure that future industry and plant-specific operating experience will be incorporated into the program such that inspection and test frequencies may be increased, as appropriate, based on test and inspection results. By RAI response dated March 28, 2011, the applicant stated that the revised Non-EQ Inaccessible Medium Voltage Cable Program specifies that cable inspection and test frequencies will be adjusted as necessary based on the results of cable testing and manhole inspections. The applicant also stated that it will incorporate lessons learned from future industry and plant-specific operating experience, including plant-specific test and inspection results during implementation of the Non-EQ Inaccessible Medium Voltage Cable Program. The adjustment of inspection and test frequencies based on inspection and test results and lessons learned during the implementation of the applicant's Non-EQ Inaccessible Medium-Voltage Cable Program is consistent with staff recommendations. The staff's concern related to the evaluation of plant-specific and industry operating experience during program implementation described in RAI 3.0.3.1.6-1 is resolved.

Conclusion. On the basis of its audit and review of the applicant's Non-EQ Inaccessible Medium-Voltage Cable Program, the staff finds: (a) the applicant's program is based on and consistent with GALL AMP XI.E3, (b) the program enhancements, including the incorporation of 400 V to 2 kV power cables, are consistent with industry operating experience and current staff recommendations. The staff concludes that the applicant has demonstrated that the effects of aging will be adequately managed so that the intended function(s) will be maintained consistent with the CLB for the period of extended operation, as required by 10 CFR 54.21(a)(3). The staff also reviewed the UFSAR supplement for this AMP and concludes that it provides an adequate summary description of the program, as required by 10 CFR 54.21(d).

3.0.3.1.10 One-Time Inspection – Small Bore Piping Program

Summary of Technical Information in the Application. The applicant provided additional information related to the One-Time Inspection – Small Bore Piping Program subsequent to the issuance of the SER. The additional information is discussed below in the "Staff Evaluation" section.

Staff Evaluation. The staff's evaluation of the applicant's proposed One-Time Inspection–Small Bore Piping Program (ASME Code Class 1) is documented in SER Section 3.0.3.1.10. The applicant provided additional information subsequent to the issuance of the SER. The staff's evaluation of the additional information related to the One-Time Inspection – Small Bore Piping Program is discussed below in the following categories: (1) Volumetric Examination Methodology; (2) One-Time Inspection of Butt Welds at IP2, and Both Butt Welds and Socket Welds at IP3; (3) Destructive Examination as an Alternative; and (4) Socket Weld Periodic Inspection at IP2.

(1) Volumetric Examination Methodology

By letter dated July 26, 2010, the applicant provided supplemental information related to the One-Time Inspection – Small Bore Piping Program regarding the volumetric examination of small bore piping. The applicant stated that it will perform the examination in accordance with guidelines established in MRP-146, "Materials Reliability Program Management of Thermal Fatigue in Normally Stagnant Non-Isolable Reactor Coolant System Branch Lines," June 2005. The staff noted that MRP-146 recommends examination of the base metal half an inch beyond the toe of the weld; however, the weld metal is not required as part of examination volume. It was not clear to the staff if the applicant's inspections as part of the One-Time Inspection – Small Bore Piping Program were to inspect the base metal only without inspecting the weld metal. The staff noted that if weld metal is not inspected, this proposed inspection methodology may not be adequate to manage age-related degradation of ASME Code Class 1 small-bore piping, because industry operating experience has indicated numerous failures in small bore piping, predominantly in the form of cracking in the weld metal. The staff noted that many of these failures are documented in Licensee Event Reports (LERs). The staff's concern is that cracking in the weld metal could occur and remain undetected if examination of the weld metal is not performed. By letter dated February 10, 2011, the staff issued RAI 3.0.3.1.10-1, Part 1 and Part 2, requesting the applicant to justify that the examination volume is sufficient and capable of detecting cracking in the welds.

In its response dated March 28, 2011, the applicant stated that volumetric examinations in accordance with MRP-146 will be performed to manage thermal fatigue. The applicant clarified that volumetric examinations of butt welds always include the weld metal and that it will also perform volumetric examinations on the weld metal for socket welds.

The staff finds the applicant's response and volumetric examination methodology acceptable because the applicant's volumetric examination includes inspection of the weld metal for butt welds and socket welds. The staff's concern described in RAI 3.0.3.1.10-1, Part 1 and 2, is resolved.

(2) One-Time Inspection of Butt Welds at IP2, and Both Butt Welds and Socket Welds at IP3

By letter dated July 26, 2010, the applicant provided its inspection sampling size related to the One-Time Inspection – Small Bore Piping Program, which states that, as part of the program, it had performed volumetric inspection on two socket welds at IP2, and on 21 socket welds at IP3. The staff noted that the applicant did not provide specific information regarding the small bore piping weld population or inspection sampling size for either butt welds or socket welds. This information is needed by the staff to evaluate the adequacy of the applicant's inspection sampling for socket welds and butt welds and whether the applicant's program is consistent with the staff's inspection sampling guidance. By letter dated February 10, 2011, the staff issued RAI 3.0.3.1.10-2 requesting the applicant to provide the total population of ASME Code Class-1 small-bore welds of each weld type (butt weld and socket weld) for each unit. In addition, the staff requested the applicant to provide the inspection sampling size for ASME Code Class 1 small-bore piping.

In its response dated March 28, 2011, the applicant provided the population for each weld type at each unit. Specifically, there are 195 butt welds and 433 socket welds at IP2; there are 96 butt welds and 333 socket welds at IP3. In addition, the applicant provided the inspection sampling size consisting of examinations already completed and examinations to be completed as part of the One-Time Inspection – Small Bore Piping Program. Specifically, the program will include six butt welds (approximately 3 percent) at IP2 and four butt welds (approximately 4 percent) and 20 socket welds (approximately 6 percent) at IP3. The applicant also stated that the inspection locations of small-bore piping welds are determined by risk-informed selection criteria which will include the most susceptible and risk-significant welds. In addition, the applicant stated that no evidence of cracking or leakage has been detected based on the examinations already completed as part of the One-Time Inspection – Small Bore Piping Program. The staff noted that both IP2 and IP3 have operated for more than 30 years without any plant-specific operating experience of butt weld failures at either unit and no socket weld failures at IP3. Therefore the staff finds that use of One-Time Inspection – Small Bore Piping Program is appropriate for IP2 in regards to butt welds, and is appropriate to IP3 in regards to both socket welds and butt welds.

Based on the applicant's plant-specific operating experience and operating history, the staff's sampling guidance is three percent of the weld population, up to 10 welds, for each weld type at each unit. The staff finds the applicant's response and inspection sampling size and methodology regarding the one-time inspection of butt welds at IP2, and both butt welds and socket welds at IP3 is acceptable. In this regard, the applicant's One-Time Inspection – Small Bore Piping Program includes (a) volumetric inspection of at least three percent of the socket welds and butt welds in each unit, and (b) a methodology to select the most susceptible and risk-significant welds, consistent with the staff's sampling guidance. Also, IP2 and IP3 have operated for more than 30 years with no plant-specific operating experience that indicates cracking or failures with regards to the components in scope of the One-Time Inspection – Small Bore Piping Program, and no evidence of cracking or leakage has been detected based on the examinations already completed as part of the One-Time Inspection – Small Bore Piping Program. The staff's evaluation of the applicant's proposed inspection of socket welds at IP2 is further discussed below in category 4 (Socket Weld Periodic Inspection at IP2). The staff's concern described in RAI 3.0.3.1.10-2 is resolved.

The staff noted that the applicant's period of extended operation begins on September 28, 2013, for IP2, and on December 12, 2015, for IP3. The staff further noted that, as part of the One-Time Inspection - Small Bore Piping Program, IP3 has already completed all of its examinations during its current (fourth) 10-year inservice inspection (ISI) interval; IP2 has completed all but one of its examinations (an examination of one butt weld) during its current (fourth) 10-year ISI interval and will complete the remaining one (butt weld examination) prior to entering the period of extended operation.

The staff finds the applicant's implementation of these one-time inspections timely and acceptable because the applicant has already completed its proposed inspections, except an examination of one butt weld at IP2, without revealing any evidence of cracking or leakage and the remaining examination will be completed prior to the period of extended operation, consistent with the staff's program implementation guidance.

(3) Destructive Examination as an Alternative

By letter dated March 28, 2011, the applicant indicated that, for socket weld examination, it proposes the option of performing destructive examination in lieu of volumetric examination on a two-for-one basis. Based on the staff's sampling guidance, an applicant may take credit for each weld destructively examined as being equivalent to having volumetrically examined two welds because more information can be obtained from a destructive examination than from nondestructive examination; therefore, the staff finds the applicant's option to perform destructive examination in lieu of volumetric examination on a two-for-one basis acceptable and consistent with the staff's sampling guidance.

(4) Socket Weld Periodic Inspection at IP2

In its review of the supplemental information provided by the applicant by letter dated July 26, 2010, the staff noted cases of ASME Code Class 1 socket weld failures at IP2. The most recent case was a socket weld failure that resulted in reactor coolant system (RCS) leakage, as documented in LER 2472010004 (ADAMS Accession No. ML1014501190). Based on this plant-specific operating experience at IP2, the staff determined that the use of a one-time inspection may not be appropriate, and that it may be necessary for the applicant to perform periodic inspection of socket welds to manage age-related degradation during the period of extended operation. By letter dated February 10, 2011, the staff issued RAI 3.0.3.1.10-1, Part 3, requesting the applicant to provide justification that periodic inspection of ASME Code Class 1 small-bore socket welds at IP2 is not needed.

In its response dated March 28, 2011, the applicant provided a discussion of its operating experience which states it has experienced five leaks during 38 years of operation at IP2. In addition, the applicant stated that it will perform periodic inspection of ASME Code Class 1 socket welds at IP2. Specifically, it will volumetrically inspect at least 10 socket welds in 2012 and at least 10 socket welds during each 10-year ISI interval. However, the basis for the inspection sampling size for the periodic inspection at IP2 was not clear to the staff. A robust inspection program for socket welds is warranted based on the operating experience at IP2, and the inspection sampling should be statistically significant so that cracking, if present, will be detected prior to leakage of the component. In the case of a focused inspection in which an applicant selects the most susceptible welds and the most risk-significant welds, the staff recommends inspection sampling of 10 percent, up to 25 welds, for each 10-year interval during the period of extended operation. Therefore, by letter dated June 15, 2011, the staff issued RAI 3.0.3.1.10-3 requesting the applicant to justify the sampling adequacy of the periodic inspection for socket welds at IP2.

In its response dated July 14, 2011, the applicant stated that it will perform periodic volumetric inspection of socket welds as part of its ISI Program. The staff's evaluation of the applicant's ISI Program is documented in Section 3.0.3.3.4 of the SER. The staff noted that the inspection sampling will consist of 25 socket welds during each of the ISI 10-year intervals, consistent with the staff's inspection sampling guidance. In addition, since IP2 is currently in the third period of its fourth ISI 10-year interval, the applicant will perform volumetric examination of seven socket welds (i.e., 28 percent of the 25 welds needed for the interval) which is the prorated number of socket welds to be inspected for the remaining years in the current 10-year interval. The staff noted that the applicant will use a methodology to select the most susceptible and risk-significant welds to ensure a high probability of detecting cracking, if it exists. The staff also noted if cracking is detected during the inspection, there will be an extent of condition review to evaluate the inspection sampling size to ensure that it is adequate to identify cracking that could occur at other locations.

The staff finds the applicant's response acceptable because the applicant's sampling methodology ensures that an adequate number of welds will be examined during each 10-year interval to ensure that aging, if present, will be adequately managed during the period of extended operation. The staff's concern described in RAI 3.0.3.1.10-1, Part 3 and RAI 3.0.3.1.10-3 is resolved.

Based on its review, the staff determined that the applicant's proposed aging management of ASME Code Class 1 small bore piping is adequate because the program includes a sufficient number of welds to be inspected, an adequate selection methodology that focuses on susceptibility, welds and risk-significance, and the program will also be implemented in a reasonable timeframe.

Operating Experience. There are no changes or updates to this section of the SER.

UFSAR Supplement. There are no changes or updates to this section of the SER.

Conclusion. There are no changes or updates to this section of the SER.

### 3.0.3.2 AMPs Consistent with the GALL Report with Exceptions and/or Enhancements

3.0.3.2.10  Masonry Wall Program

Summary of Technical Information in the Application. The applicant provided additional information related to the Masonry Wall Program subsequent to the issuance of the SER. The additional information is discussed below in the "Staff Evaluation" section.

Staff Evaluation. The staff's evaluation of the applicant's proposed Masonry Wall Program is documented in Section 3.0.3.2.10 of the SER.

Subsequent to issuance of the SER, the staff identified a need for additional information regarding the frequency of inspections for masonry walls within the scope of license renewal. By letter dated February 10, 2011, the staff issued RAI 3.0.3.2.10-1 requesting the applicant to provide the inspection interval for masonry walls.

By letter dated March 28, 2011, the applicant responded and stated that the inspection interval for masonry walls within the scope of license renewal is every 5 years. The staff reviewed the applicant's response and found it acceptable because the applicant is conducting inspections consistent with the frequency described in the GALL Report.

Operating Experience. There are no changes or updates to this section of the SER.

UFSAR Supplement. There are no changes or updates to this section of the SER.

Conclusion. There are no changes or updates to this section of the SER.

3.0.3.2.15  Structures Monitoring Program

Summary of Technical Information in the Application. The applicant provided additional information related to the Structures Monitoring Program subsequent to the issuance of the SER. The additional information is discussed below in the "Staff Evaluation" section.

Staff Evaluation. The staff's evaluation of the applicant's proposed Structures Monitoring Program is documented in Section 3.0.3.2.15 of the SER. Subsequently, the staff requested additional information regarding the Structures Monitoring Program acceptance criteria. The staff's evaluation of the additional information submitted by the applicant related to the Structures Monitoring Program is discussed below.

GALL AMP XI.S6, "Structures Monitoring Program," states that American Concrete Institute (ACI) 349.3R is an acceptable basis for selection of parameters monitored, detection of aging effects, and acceptance criteria. The LRA states that the applicant's program incorporates inspection guidance based on recommendations contained in ACI 349.3R; however, the LRA does not clearly state that acceptance criteria align with those in ACI 349.3R. By letter dated

February 10, 2011, the staff issued RAI 3.0.3.2.15-1 requesting the applicant provide quantitative acceptance criteria that align with ACI 349.3R, or provide technical justification for any differences.

By letter dated March 28, 2011, the applicant responded and stated that its Structures Monitoring Program has a responsible engineer with the appropriate education and experience to identify and evaluate existing conditions using appropriate standards, including ACI standards. The applicant further stated that the program will be enhanced to include more detailed guidance on quantitative acceptance criteria of ACI 349.3R, "Evaluation of Existing Nuclear Safety-Related Concrete Structures." The applicant committed to implementing this enhancement prior to the period of extended operation (Commitment 25).

The staff reviewed the applicant's response and found it acceptable because the applicant has committed to enhance its Structures Monitoring Program to include acceptance criteria aligned with the quantitative criteria recommended in ACI 349.3R, and therefore, with the recommendations in the GALL Report. The applicant has committed to implement this enhancement prior to the period of extended operation.

Operating Experience. There are no changes or updates to this section of the SER.

UFSAR Supplement. In LRA Section A.2.1.35, the applicant provided the UFSAR supplement for the Structures Monitoring Program. In a letter dated March 28, 2011, the applicant supplemented the application and revised Commitment No. 25 to enhance the acceptance criteria of the program. The staff has determined that the information in the UFSAR supplement is an adequate summary description of the program, as required by 10 CFR 54.21(d).

Conclusion. There are no changes or updates to this section of the SER.

### 3.0.3.3 AMPs Not Consistent with or Not Addressed in the GALL Report

3.0.3.3.1 Boral Surveillance Program

Summary of Technical Information in the Application. The applicant provided additional information related to the Boral Surveillance Program subsequent to the issuance of the SER. The additional information is discussed below in the "Staff Evaluation" section.

Staff Evaluation. The staff's evaluation of the applicant's Boral Surveillance Program is documented in Section 3.0.3.3.1 of the SER. Subsequent to the issuance of the SER, the staff identified a need for additional information as to whether the Boral Surveillance Program will perform inspection and testing activities with sufficient frequency to ensure that aging of Boral neutron-absorbing panels will be adequately managed. By letter dated June 15, 2011, the staff issued RAI B.1.4-1 requesting that the applicant state the frequency of inspection and testing activities, provide justification if the frequency is less than once every 10 years, and revise the UFSAR supplement to include the inspection and testing interval.

In its response dated July 14, 2011, the applicant stated that inspection and testing activities are based on plant-specific operating experience and will occur at least once every 10 years during the period of extended operation. The applicant revised the UFSAR supplement to reflect the minimum 10-year frequency. The staff finds the applicant's response acceptable because the applicant's inspection and testing of Boral coupons are informed by operating experience and will be performed at least once every 10 years, which is sufficient to ensure that Boral degradation will be detected prior to loss of intended function. The staff's concern described in RAI B.1.4-1 is resolved.

UFSAR Supplement. There are no changes or updates to this section of the SER.

Conclusion. There are no changes or updates to this section of the SER.

3.0.3.3.4 Inservice Inspection Program

Summary of Technical Information in the Application. There are no changes or updates to this section of the SER.

Safety Evaluation. There are no changes or updates to this section of the SER.

Operating Experience. There are no changes or updates to this section of the SER.

UFSAR Supplement. In LRA Section A.2.1.17, the applicant provided the UFSAR supplement for the Inservice Inspection Program. By letter dated July 14, 2011, the applicant revised LRA Section A.2.1.17 to include the following,

> IPEC will perform twenty-five volumetric weld metal inspections of small-bore Class 1 socket welds during each 10-year ISI interval scheduled as specified by IWB-2412 of the ASME Section XI Code. In lieu of volumetric examinations, destructive examinations may be performed, where one destructive examination may be substituted for two volumetric examinations.

The applicant committed (Commitment 46) to perform 25 volumetric weld metal inspections of small bore Class 1 socket welds in each 10-year ISI interval during the period of extended operation.

The staff reviewed LRA Section A.2.1.17 as amended by letter dated July 14, 2011, and concludes that this section of the UFSAR supplement provides an adequate summary description of the program, as required by 10 CFR 54.21(d).

Conclusion. There are no changes or updates to this section of the SER.

### 3.0.5  Operating Experience for Aging Management Programs

#### 3.0.5.1  Summary of Technical Information in the Application

LRA Section B.0.4 describes the consideration of operating experience for AMPs. The LRA states that the applicant reviewed past operating experience to prepare the application. This review included operating experience from such sources as corrective actions; reports of recent inspections, examinations, tests, and sample results; input from program owners; and applicable self-assessments, quality assurance (QA) audits, peer evaluations, and NRC reviews. The LRA also states that site procedures require reviews of site and relevant industry operating experience as the site continues operation through the period of extended operation.

### 3.0.5.2    *Staff Evaluation*

Pursuant to 10 CFR 54.21(a)(3), an applicant is required to demonstrate that the effects of aging on structures and components (SCs) subject to an aging management review (AMR) will be adequately managed so that their intended functions will be maintained consistent with the CLB for the period of extended operation. SRP-LR, Revision 1, Section A.1.2.3 describes 10 elements of an acceptable AMP including element 10, "Operating Experience."

The staff reviewed LRA Section B.0.4 to determine whether the applicant will implement adequate programmatic activities for the continual review of both plant-specific and industry operating experience to identify areas where AMPs should be enhanced or new AMPs developed. While LRA Section B.0.4 states that operating experience will be reviewed in the future, it does not fully describe the details of how the applicant will use future operating experience to ensure that the AMPs will remain effective for managing the aging effects during the period of extended operation. Also, it is not clear as to which AMPs will be updated based on future operating experience or whether new AMPs will be developed, as necessary. By letter dated June 15, 2011, the staff issued RAI B.0.4-1 requesting that the applicant describe in detail the programmatic activities that will be used to continually identify aging issues, evaluate them, and, as necessary, enhance the AMPs or develop new AMPs.

In its response dated July 14, 2011, the applicant provided further information to describe how it will use its existing programs to monitor, on an ongoing basis, plant-specific and industry operating experience and how the evaluations completed under these programs will ensure that the AMPs will be effective in managing the aging effects for which they are credited. The applicant stated that it will use two programs: the operating experience program (OEP) and the corrective action program (CAP). The applicant indicated that the OEP will monitor sources of industry operating experience whereas the CAP will monitor sources of plant-specific operating experience. Some examples of industry sources are NRC generic communications and Institute of Nuclear Power Operations event reports, and some plant-specific sources are the results of inspections performed under plant programs and system health reports. The applicant stated that items are first screened under the OEP to determine the potential impact to the plants, and are entered into the CAP when the item concerns degraded equipment. The applicant explained that degraded equipment includes degradation due to the effects of aging. Evaluations under the CAP consider whether the frequency of future inspections needs adjustment, whether new or different inspections are needed, and whether inspections include adequate depth and breadth of component, material, and environment combinations. The applicant stated that corrective actions can include enhancement of existing AMPs or development of new AMPs. Finally, the applicant stated that both the OEP and the CAP are administratively controlled.

The staff finds the applicant's response acceptable because the programmatic activities it described are adequate to monitor and evaluate plant-specific and industry operating experience on a continual basis. In addition, these programs provide for the enhancement of AMPs or the development of new AMPs, when necessary, to ensure that the effects of aging will be adequately managed. The staff's concern described in RAI B.0.4-1 is resolved.

The staff determines that the applicant has met the intent of the "operating experience" program element with respect to the future consideration of operating experience.

### 3.0.5.3    UFSAR Supplement

The staff reviewed the UFSAR supplements in LRA Appendix A to determine whether the applicant provided an adequate summary description of the ongoing operating experience review activities.  As the staff found no such description, it also requested in RAI B.0.4-1 that the applicant provide a description of these activities for the UFSAR supplement required by 10 CFR 54.21(d).

In its response dated July 14, 2011, the applicant did not amend the UFSAR supplements to include a description of the programmatic activities for the ongoing review of operating experience.  As such, on July 21, 2011, staff held a teleconference with the applicant to discuss the need for such a description in the UFSAR supplements.  On July 27, 2011, the applicant submitted a supplemental response to RAI B.0.4-1, in which it provided the following description in both UFSAR supplements:

> The Operating Experience Program (OEP) and the Corrective Action Program (CAP) help to assure continued effectiveness of aging management programs through evaluations of operating experience.  The OEP implements the requirements of NRC NUREG-0737, "Clarification of TMI Action Plan Requirements," Section I.C.5 and evaluates site, Enterqy fleet, and industry operating experience for impact on IPEC.  The CAP implements the requirements of 10 CFR 50, Appendix B, Criterion XVI and is used to evaluate and effect appropriate actions in response to operating experience relevant to IPEC that indicates a condition adverse to quality or a non-conformance.

The staff reviewed this description against the acceptance criteria in SRP-LR Sections 3.1.2.4, 3.2.2.4, 3.3.2.4, 3.4.2.4, 3.5.2.4, and 3.6.2.4.  In accordance with the acceptance criteria in these sections, the staff determines that the summary description is sufficiently comprehensive such that later changes can be controlled by 10 CFR 50.59.  As such, the staff determines that the information in the UFSAR supplements is an adequate summary description of the ongoing operating experience review activities, as required by 10 CFR 54.21(d).

### 3.0.5.4    Conclusion

On the basis of its review of the applicant's programmatic activities for the ongoing review of operating experience, the staff concludes that the applicant has demonstrated that the effects of aging will be adequately managed so that the intended function(s) will be maintained consistent with the CLB for the period of extended operation, as required by 10 CFR 54.21(a)(3).  The staff also reviewed the UFSAR supplement for these activities and concludes that it provides an adequate summary description, as required by 10 CFR 54.21(d).

## 3.1  Aging Management of Reactor Vessel, Internals and Reactor Coolant Systems

### 3.1.2  Staff Evaluation

### 3.1.2.1  AMR Results Consistent with the GALL Report

LRA Table 3.1.1, item number 82, states that SG primary side divider plates are composed of nickel alloy. LRA Table 3.1.1, item number 81, addresses cracking due to primary water stress corrosion cracking (PWSCC) for the nickel alloy or nickel-alloy clad SG divider plate exposed to reactor coolant. The applicant credited the Water Chemistry Control – Primary and Secondary Program to manage cracking due to PWSCC in nickel-alloy SG divider plates exposed to reactor coolant, consistent with the GALL Report.

The staff noted that, from foreign operating experience in steam generators with similar design to that of the applicant's SGs, cracking due to PWSCC has been identified in SG divider plate assemblies fabricated from Alloy 600, even with proper primary water chemistry. The staff noted specifically, that cracks have been detected in the stub runner, very close to the tubesheet/stub runner weld and with depths of almost a quarter of the divider plate's thickness. Therefore, the staff determined that the Water Chemistry – Primary and Secondary Program might not be effective in managing cracking due to PWSCC in SG divider plate assemblies fabricated from Alloy 600 and its associated weld metals.

The staff noted that, although these SG divider plate assembly cracks might not have a significant safety impact in and of themselves, these cracks could affect adjacent items that are part of the reactor coolant pressure boundary, such as the tubesheet and the channel head, if they propagate to the boundary with these items. The staff further noted that for the tubesheet, PWSCC cracks in the divider plate assemblies fabricated from Alloy 600 and its associated weld metals could propagate to the tubesheet cladding, with possible consequences to the integrity of the tube-to-tubesheet welds. Furthermore, for the channel head, the PWSCC cracks in the divider plate assemblies could propagate to the SG triple point (i.e. the point where the divider plate and tube sheet meet with the shell) and potentially affect the pressure boundary of the SG channel head.

The staff reviewed the applicant's UFSAR and noted that IP2 UFSAR Section 4.2.2.3 and Table 4.2-1, and IP3 UFSAR Section 4.2.2 and Table 4.2-1 describe the construction materials for the IP2 replacement Model 44F steam generators and for the IP3 replacement Model 44F steam generators, respectively. However, there was no information about the construction materials of the divider plate assemblies for the SGs at both units.

By letter dated February 10, 2011, the staff issued RAI 3.1.2.2.13-1 requesting that the applicant (1) discuss the materials of construction for the IP2 and IP3 SG divider plate assemblies, including the welds within these assemblies, to the channel head and to the tubesheet, (2) if any constitutive/weld material of the SG divider plate assemblies is susceptible to cracking (e.g., Alloy 600 or its associated weld materials), describe an inspection program (examination technique and frequency) to ensure that there are no cracks propagating into other items which are part of the reactor coolant pressure boundary (e.g., tubesheet and channel head) that could challenge the integrity of those adjacent items.

In its response dated March 28, 2011, the applicant described that for IP2 and IP3, the divider plates are Inconel 600 (ASME-SB-168) and that it is conservatively assumed that the weld materials are the associated Alloy 600 weld materials. The applicant further clarified that IP2 original Westinghouse Model 44 steam generators were replaced with Model 44F steam generators in 2000 and that IP3 original Westinghouse Model 44 steam generators were replaced with Model 44F steam generators in 1989. The applicant also described the evaluation and conclusion from the Electric Power Research Institute (EPRI) about the safety concern of a cracked divider plate in a Westinghouse Model F SG. In addition, the applicant described that the industry plans to study the potential for divider plate crack growth and to develop a resolution to the concern through the EPRI Steam Generator Management Program (SGMP) Engineering and Regulatory Technical Advisory Group, which is expected to be completed by 2013. However, recognizing that the EPRI SGMP resolution of this issue is under development, the applicant stated that it would inspect all its SGs to assess the condition of the divider plate assembly.

The applicant's RAI response dated March 28, 2011, was subsequently revised by letters dated July 14 and July 27, 2011, which considered information discussed in conference calls held on June 9 and July 25, 2011. In the letter of July 27, 2011, the applicant committed (Commitment 41) to the following:

> IPEC will perform an inspection of steam generators for both units to assess the condition of the divider plate assembly. The examination technique used will be capable of detecting PWSCC in the steam generator divider plate assemblies. The IP2 steam generator divider plate inspections will be completed within the first ten years of the period of extended operation (PEO), i.e. prior to September 28, 2023. The IP3 steam generator divider plate inspections will be completed within the first refueling outage following the beginning of the PEO.

Based on its review, the staff finds the applicant's response to RAI 3.1.2.2.13-1 and associated Commitment 41 acceptable because the applicant will assess the condition of the divider plate assembly in each SG at both units by inspection during the period of extended operation, in a time period consistent with the detection of potential PWSCC cracks, with appropriate examination techniques. The staff's concern described in RAI 3.1.2.2.13-1 is resolved.

The staff concludes that the applicant has demonstrated that the effects of aging for these components will be adequately managed so that their intended functions will be maintained consistent with the CLB during the period of extended operation, as required by 10 CFR 54.21(a)(3).

### 3.1.2.2 AMR Results Consistent with the GALL Report for Which Further Evaluation is Recommended

By letter dated June 14, 2010, the applicant provided LRA amendment number 9, which included the Reactor Vessel Internals Program and revisions to the related LRA Sections. The LRA amendment described the Reactor Vessel Internals Program as a new, plant-specific aging management program. The Reactor Vessel Internals Program and associated LRA revisions were based on the Electric Power Reactor Institute (EPRI) report, "Materials Reliability Program: Pressurized Water Reactor Internals Inspection and Evaluation Guidelines (MRP-227, Rev. 0)," which was submitted to the NRC by letter dated January 12, 2009 (ADAMS Accession No. ML090160204). The staff completed its review of MRP-227, Rev. 0, and issued its SER on that report on June 22, 2011 (ADAMS Accession No. ML111600498). Subsequent to issuing its SER on MRP-227, Rev. 0, the staff issued Regulatory Information Summary (RIS) 2011-07, "License Renewal Submittal Information for Pressurized Water Reactor Internals Aging Management." As stated therein, RIS 2011-07 "provides information to licensees with respect to how to meet their existing license renewal commitments related to reactor vessel internals aging management programs, and on acceptable changes to existing license renewal commitments in order to account for recent issue of the staff's final Safety Evaluation (SE) on MRP-227, Rev. 0, and the forthcoming issue of the approved version of MRP-227."

RIS 2011-07 identifies categories for plants in varying stages of license renewal. Category "C" is described as being applicable to plants that have an LRA currently under review which applies to IP2 and IP3. RIS 2011-07 states, for applicants in Category C, that "Applicants will be expected to revise their commitment for aging management of PWR vessel internals such that the submittal information identified in the SE for MRP-227 would be submitted to the NRC for review and approval not later than two years after issuance of the renewed license and not later than two years before the plant enters the period of extended operation, whichever comes first." The applicant's current licenses expire on September 28, 2013 for IP2 and December 12, 2015 for IP3.

By letter dated August 22, 2011, the applicant stated that it would further supplement its LRA by September 28, 2011, to include an inspection plan for reactor vessel internals, which is consistent with its previous commitment (Commitment 30). In addition, the applicant stated that following the issuance of the approved version of MRP-227 (MRP-227-A), it will review the inspection plan to determine any need for revision, and will modify the inspection plan to include the necessary revisions, if any. The submission of the Reactor Vessel Internals Program, along with applicant's stated intention of supplying an inspection plan by September 28, 2011, and its plan to revise its program, as necessary, to address the staff's evaluation of MRP-227, is consistent with the applicant's previous commitment and the recent guidance provided in RIS 2011-07. Therefore, the staff's conclusions regarding the applicant's AMRs for reactor vessel internals, as documented in SER Sections 3.1.2.2.6, 3.1.2.2.9, 3.1.2.2.12, 3.1.2.2.15, and 3.2.2.17, remain valid.

3.1.2.2.16  Cracking Due to Stress Corrosion Cracking and Primary Water Stress Corrosion Cracking

SER Section 3.1.2.2.16 discussed (1) cracking due to stress corrosion cracking (SCC) in stainless steel control rod drive head penetration components and on the primary coolant side of steel steam generator heads clad with stainless steel, and (2) cracking due to SCC that could occur on stainless steel pressurizer spray heads and cracking due to PWSCC that could occur on nickel alloy pressurizer spray heads. This SSER supplements the discussion in SER Section 3.1.2.2.16(1) which appears on pages 3-286 through 3-289 of the SER. SER Section 3.1.2.2.16(2) is unchanged by this SSER.

(1)  Subsequent to the issuance of the SER, the staff requested additional information related to the aging management of SG tube-to-tubesheet welds made or clad with nickel alloy. The following is the staff's evaluation of the additional information.

LRA Table 3.1.1, item number 3.1.1-35, states that the corresponding GALL Report line item applies to once-through steam generators (SGs) and was used as a comparison for the SG tubesheets at the applicant's site. The applicant further stated that for the steel with nickel alloy clad steam generator tubesheets, cracking is managed by the Water Chemistry Control – Primary and Secondary and Steam Generator Integrity Programs.

SRP-LR Section 3.1.2.2.16 identifies that cracking due to PWSCC could occur on the primary coolant side of PWR steel SG tube-to-tubesheet welds made or clad with nickel alloy. The GALL Report recommends ASME Section XI ISI Program and control of water chemistry to manage cracking due to PWSCC and recommends no further aging management review for PWSCC of nickel alloy if the applicant complies with applicable NRC Orders and provides a commitment in the FSAR supplement to implement applicable (1) Bulletins and Generic Letters and (2) staff-accepted industry guidelines. Item IV.D2-4 in the GALL Report addresses cracking due to PWSCC, and is applicable only to once-through SGs, but not to recirculating SGs.

The staff noted that ASME Code, Section XI, does not require any inspection of the tube-to-tubesheet welds, and that no specific NRC Orders or bulletins address inspection requirements for these welds. The staff's concern is that, if the tubesheet cladding is Alloy 600 or the associated Alloy 600 weld materials, the region of the autogenous tube-to-tubesheet weld may have insufficient chromium content to prevent initiation of PWSCC, even when the SG tubes are made from Alloy 690TT. Consequently, a crack initiated in this region, close to a tube, may propagate into or through the weld, causing a failure of the weld and of the reactor coolant pressure boundary (RCPB). This could occur in once-through SGs, as well as in recirculating SGs such as those used at both of the applicant's units. Therefore, unless the NRC has approved a redefinition of the RCPB in which the autogenous tube-to-tubesheet weld is no longer included, or the tubesheet cladding and welds are not susceptible to PWSCC, the staff considers that the effectiveness of the Water Chemistry Control – Primary and Secondary Program should be verified to ensure PWSCC cracking is not occurring. Moreover, it was not clear to the staff how the Steam Generator Integrity Program is able to manage PWSCC of the tubesheet cladding, including the tube-to-tubesheet welds.

LRA Section 2.3.1.4 describes the IP2 replacement Westinghouse Model 44F SG tubes as being fabricated from Alloy 600TT, and the IP3 replacement Westinghouse Model 44F SG tubes as being fabricated from Alloy 690TT. The applicant also described the tubesheet surfaces in contact with reactor coolant as clad with Inconel, and stated that the tube-to-tubesheet joints are welded for both units' SGs.

By letter dated February 10, 2011, the staff issued RAI 3.1.2.2.16-1 requesting that the applicant, for the IP2 SGs, (1) clarify whether the tube-to-tubesheet welds are included in the RCPB or alternate repair criteria have been permanently approved, and (2) if the SGs do not have permanently approved alternate repair criteria, justify how the Steam Generator Integrity Program is capable of managing PWSCC in tube-to-tubesheet welds, or provide a plant-specific AMP that will complement the Water Chemistry Control – Primary and Secondary Program, in order to verify its effectiveness and ensure that cracking due to PWSCC is not occurring in tube-to-tubesheet welds. For the IP3 SGs, the staff requested that the applicant justify how the Steam Generator Integrity Program is capable of managing PWSCC in tube-to-tubesheet welds, or provide either a plant-specific AMP that will complement the Water Chemistry Control – Primary and Secondary Program, in order to verify its effectiveness and ensure that cracking due to PWSCC is not occurring in tube-to-tubesheet welds, or a rationale for why such a program is not needed.

In its response dated March 28, 2011, the applicant clarified that in the IP2 SGs, the tube-to-tubesheet welds are included in the RCPB and no tubesheet region alternate repair criterion is employed. The applicant further stated that, for each unit, it would address the potential failure of the steam generator reactor coolant pressure boundary due to PWSCC cracking of tube-to-tubesheet welds via one of two options, an analysis or an inspection. The applicant further stated that an approved analytical evaluation would obviate the need to develop a plant-specific AMP to verify effectiveness of the Water Chemistry Control – Primary and Secondary program.

The applicant's RAI response dated March 28, 2011, was subsequently revised by letters dated July 14, July 27, and August 9, 2011, which considered, among other things, information discussed in conference calls held on June 9, and July 25, 2011. In the August 9, 2011 letter, the applicant committed (Commitment 42) to the following:

IPEC will develop a plan for each unit to address the potential for cracking of the primary to secondary pressure boundary due to PWSCC of tube-to-tubesheet welds using one of the following two options.

Option 1 (Analysis)

IPEC will perform an analytical evaluation of the steam generator tube-to-tubesheet welds in order to establish a technical basis for either determining that the tubesheet cladding and welds are not susceptible to PWSCC, or redefining the pressure boundary in which the tube-to-tubesheet weld is no longer included and, therefore, is not required for reactor coolant pressure boundary function. The redefinition of the reactor coolant pressure boundary must be approved by the NRC as part of a license amendment request.

Option 2 (Inspection)

IPEC will perform a one-time inspection of a representative number of tube-to-tubesheet welds in each steam generator to determine if PWSCC cracking is present. If weld cracking is identified:

a. The condition will be resolved through repair or engineering evaluation to justify continued service, as appropriate, and
b. An ongoing monitoring program will be established to perform routine tube-to tubesheet weld inspections for the remaining life of the steam generators.

Moreover, for IP2, the applicant stated that the tube-to-tubesheet welds have been in service for approximately eleven years since it replaced the IP2 SGs in 2000. Considering this limited service time, the applicant further stated that, if Option 1 were not implemented, it would implement Option 2 that includes tube-to-tubesheet weld inspections for PWSCC. The applicant further stated that these inspections would be performed between March 2020 and March 2024, such that the SGs will have been in service between 20 and 24 years. For IP3, the applicant stated that the tube-to-tubesheet welds have been in service for approximately 22 years since it replaced the IP3 SGs in 1989. The applicant further stated that, if Option 1 were not implemented, it would implement Option 2 which includes tube-to-tubesheet weld inspections for PWSCC. The applicant further stated that these inspections would be performed prior to the end of the first refueling outage following the beginning of the period of extended operation.

Based on its review, the staff finds the applicant's response to RAI 3.1.2.2.16-1 and associated Commitment 42 acceptable because the applicant will manage the aging effect of cracking due to PWSCC in the SG tube-to-tubesheet welds either by demonstrating that those welds are no longer included in the SG reactor coolant pressure boundary function or are not susceptible to PWSCC, or by implementing a one-time inspection on a representative number of tube-to-tubesheet welds of each steam generator to determine if PWSCC is present, in a time period consistent with the detection of potential PWSCC cracks. The staff finds that the timing of this inspection for each unit is acceptable because at the time of the inspections, the respective SGs will have been in operation for between 20 and 24 years, and between 22 and 28 years, for IP2 and IP3, respectively, and it is unlikely that significant detrimental PWSCC cracking will have initiated at this time period. The staff also noted that if aging effects are identified by the inspections, the applicant will take corrective actions including an evaluation of the degradation and the implementation of routine inspections of the tube-to-tubesheet welds for the remaining life of the SGs. The staff's concern described in RAI 3.1.2.2.16-1 is resolved.

## 3.2 Aging Management of Engineered Safety Features Systems

### 3.2.2 Staff Evaluation

### 3.2.2.3 AMR Results Not Consistent with or Not Addressed in the GALL Report

3.2A.2.3.5  Containment Penetrations-Summary of Aging Management Review-LRA Table 3.2.2-5-IP2

SER Section 3.2A.2.3.5 presented the staff's review of AMR items in LRA Table 3.2.2-5-IP2. In addition to the AMR results documented in the SER for LRA Table 3.2.2-5-IP2, by letter dated July 26, 2010, the applicant proposed no aging effect for stainless steel filter housings exposed to indoor air externally. This line item is similar to Item VF-12 in the GALL Report, which is for stainless steel piping, piping components, and piping elements in an external environment of air—indoor uncontrolled. Because the LRA item is similar to the GALL Report item for that material and environment, the staff finds that the exposure of stainless steel material to plant indoor air will not result in aging that will be of concern during the period of extended operation.

On the basis of its review, the staff finds that the applicant has appropriately evaluated the AMR results of material, environment, aging effect requiring management and AMP combinations not addressed in the GALL Report. The staff finds that the applicant has demonstrated that the effects of aging will be adequately managed so that the intended functions will be maintained consistent with the CLB for the period of extended operation, as required by 10 CFR 54.21(a)(3).

## 3.3 Aging Management of Auxiliary Systems

### 3.3.2 Staff Evaluation

### 3.3.2.3 AMR Results Not Consistent with or Not Addressed in the GALL Report

**3.3B.2.3.19 Chlorination System, Nonsafety-Related Components Potentially Affecting Safety Functions – Summary of Aging Management Review – LRA Table 3.3.2-19-5-IP3**

SER Section 3.3B.2.3.19 presented the staff's review of AMR items in LRA Table 3.2.2-19-5-IP2. In addition to the AMR results documented in the SER for LRA Table 3.3.2-19-5-IP3, by letter dated July 26, 2010, the applicant proposed no aging effect for plastic piping exposed to indoor air externally and treated water internally. The staff noted that the applicant's LRA was submitted using GALL Report Revision 1 which did not address this component, material and environment combination. The staff also noted that GALL Report Revision 2 addresses polyvinyl chloride (PVC) pipe exposed to the air-indoor environment and condensation. No aging effects for PVC in these environments would be expected based on GALL items AP-268, SP-152, and SP-153 which state in part that generally low operating temperatures and historical good chemical resistance data for PVC components, combined with a lack of historic negative operating experience, indicate that PVC is not likely to experience any degradation from the nonaggressive indoor air and condensation. The staff noted that the internal environment would include chlorine during chemical injection periods. The staff also noted that the Environmental Technical Specifications and UFSAR state a range of chlorine from 13.5 percent to 15 percent. The staff further noted that the "Chemical Resistance of Plastics and Elastomers," Third Edition, by the Plastic Design Library Staff, William Andrew Publishing, Interactive Table, states that PVC material exposed to sodium hypochlorite solutions up to 20 percent are acceptable with no effect. Therefore the staff finds the applicant's AMR results as appropriate.

On the basis of its review, the staff finds that the applicant has appropriately evaluated the AMR results of material, environment, aging effect requiring management and AMP combinations not addressed in the GALL Report. The staff finds that the applicant has demonstrated that the effects of aging will be adequately managed so that the intended functions will be maintained consistent with the CLB for the period of extended operation, as required by 10 CFR 54.21(a)(3).

THIS PAGE INTENTIONALLY LEFT BLANK

# SECTION 4

# TIME LIMITED AGING ANALYSIS

## 4.3 Metal Fatigue Analyses

## 4.3.3 Effects of Reactor Water Environment on Fatigue Life

### 4.3.3.1 Summary of Technical Information in the Application

The applicant provided additional information related to the Metal Fatigue Analyses subsequent to the issuance of the Safety Evaluation Report (SER). The additional information is discussed below in Supplemental SER Section 4.3.3.2 "Staff Evaluation."

### 4.3.3.2 Staff Evaluation

The staff's evaluation of the applicant's environmentally-assisted fatigue evaluations are documented in Section 4.3.3.2 of the SER. Subsequently, based on recent staff review it was noted that the applicant's plant-specific configuration may contain locations that require an analysis to determine the effects of reactor water environment on component fatigue life, other than those generic locations identified in NUREG/CR-6260 "Application of NUREG/CR-5999 Interim Fatigue Curves to Selected Nuclear Power Plant Components," February 1995. The staff's evaluation of the additional information submitted by the applicant in relation to environmentally-assisted fatigue is discussed below.

By letter dated February 10, 2010, the staff issued request for additional information (RAI) RCS-3 requesting the applicant to confirm and justify that the locations selected for environmentally assisted fatigue analyses, consistent with NUREG/CR-6260, are the most limiting for the plant. Furthermore, if these locations are not the most limiting for the plant, the applicant was requested to clarify the locations that require an environmentally-assisted fatigue analysis and the actions that will be taken for these additional locations.

In its response dated March 28, 2010, the applicant committed (Commitment 43) to implement, prior to entering the period of extended operation, the following:

> IPEC (Indian Point Energy Center) will review design basis ASME (American Society of Mechanical Engineers) Code Class 1 fatigue evaluations to determine whether the NUREG/CR-6260 locations that have been evaluated for the effects of reactor coolant environment on fatigue usage are the limiting locations for the IP2 and IP3 (Indian Point Nuclear Generating Unit Nos. 2 and 3) configuration. If more limiting locations are identified, the most limiting location will be evaluated for the effects of the reactor coolant environment on fatigue usage.

> IPEC will use the NUREG/CR6909 methodology in the evaluation of the limiting locations consisting of nickel alloy, if any.

The staff finds that the use of NUREG/CR-6909, "Effect of LWR (light water reactor) Coolant Environments on the Fatigue Life of Reactor Materials," for nickel alloy materials is acceptable because it incorporates the most recent fatigue data for determining the environmental fatigue life correction ($F_{en}$) factor for nickel alloys.

Based on its review, the staff finds the applicant's response to RAI RCS-3 and Commitment 43 acceptable because (1) the applicant will review its design basis ASME Code Class 1 fatigue evaluations to determine whether the NUREG/CR-6260 locations are the limiting components for the applicant's design, (2) if more limiting locations are identified, the applicant will perform environmentally assisted fatigue analyses for the most limiting location, (3) the applicant will use the methodology consistent with NUREG/CR-6909 in the evaluation if the limiting location identified consists of nickel alloy, (4) the applicant will complete this review prior to entering the period of extended operation, and (5) Commitment 43 is consistent with the recommendations in "Standard Review Plan for Review of License Renewal Applications for Nuclear Power Plants" (SRP-LR) Sections 4.3.2.2 and 4.3.3.2, and NUREG-1801, "Generic Aging Lessons Learned (GALL) Report," aging management program (AMP) X.M1, to consider environmental effects for the NUREG/CR-6260 locations, at a minimum. The staff's concern in RAI RCS-3 is resolved.

Separate from the staff's license renewal reviews, but as part of its review of the AP1000 design certification application, the NRC staff identified specific concerns with the use of the NB-3600 option for the computer program WESTEMS™. These concerns are described in the staff's safety evaluation and audit reports related to the AP1000 review (ADAMS Accession Nos. ML103430502 and ML110250634, respectively). Based on this documented concern, the applicant submitted a letter dated March 28, 2011, in which it provided two commitments regarding its use of the computer program WESTEMS™ for license renewal.

First, the applicant committed (Commitment 44) to include a written explanation and justification of any user intervention in future evaluations using the WESTEMS™ "Design CUF (cumulative usage factor)" module. The applicant stated that this commitment will be implemented prior to the end of the current licensing term, which is September 2013 for IP2 and December 2015 for IP3. The staff finds the applicant's implementation schedule reasonable because the applicant is ensuring that a written explanation and justification of any user intervention, in future calculations using the WESTEMS™ "Design CUF" module, will be documented. The staff noted that the implementation schedule also allows the applicant sufficient time to document and implement necessary procedures.

Based on its review, the staff finds the applicant's Commitment 44 acceptable because it ensures that the records of any calculations performed with WESTEMS™ "Design CUF" module will contain sufficient information to document and justify any assumptions and engineering judgment used to calculate the CUF value, and that the basis for the conclusions in the fatigue calculations are auditable and retrievable.

Second, in its letter of March 28, 2011, the applicant also provided Commitment 45, which states that it will not use the NB-3600 option of the WESTEMS™ program in future fatigue design calculations until the issues identified in NRC's review of the NB-3600 option of the program have been resolved. Further, this commitment will be implemented prior to the end of the current licensing term.  The staff finds the applicant's implementation schedule reasonable because the applicant is ensuring that the NB-3600 option of the WESTEMS™ program will not be used for design calculations.  The staff noted that the implementation schedule also allows the applicant sufficient time to document and implement necessary procedures to prevent the use of the NB-3600 option of the WESTEMS™ program.  The staff noted that the applicant's current decision to not use this option of the WESTEMS™ program is acceptable in view of the staff's identification of an issue regarding whether the NB-3600 option performs fatigue calculations consistent with ASME Code Section III, Subsection NB, Subarticle NB-3600.

Based on its review, the staff finds the applicant's Commitment 45 acceptable because the applicant committed not to use the NB-3600 option of the WESTEMS™ program in future design calculations until the NRC-identified issue is resolved.

### 4.3.3.3 UFSAR Supplement

There are no changes or updates to this section of the SER.

### 4.3.3.4 Conclusion

There are no changes or updates to this section of the SER.

THIS PAGE INTENTIONALLY LEFT BLANK

# SECTION 5

# REVIEW BY THE ADVISORY COMMITTEE ON REACTOR SAFEGUARDS

The staff has provided the Advisory Committee on Reactor Safeguards with a copy of this supplemental safety evaluation report.

THIS PAGE INTENTIONALLY LEFT BLANK

# SECTION 6

# CONCLUSION

The staff concludes that the additional information provided by Entergy Nuclear Operations, Inc., does not alter the conclusions stated in the SER and that the requirements of 10 CFR 54.29(a) have been met.

THIS PAGE INTENTIONALLY LEFT BLANK

# APPENDIX A

## IP LICENSE RENEWAL COMMITMENTS

During the review of the Indian Point Nuclear Generating Unit Nos. 2 and 3 (IP2 and IP3), license renewal application (LRA) by the staff of the U.S. Nuclear Regulatory Commission (NRC) (the staff), Entergy Nuclear Operations, Inc. (Entergy or the applicant) made commitments related to aging management programs (AMPs) to manage the aging effects for certain structures and components during the period of extended operation. The following table lists these commitments along with the applicant's stated implementation schedules and sources for each commitment. This list supersedes the list published in Appendix A of NUREG-1930.

| | APPENDIX A: INDIAN POINT NUCLEAR GENERATING UNIT NOS. 2 AND 3 LICENSE RENEWAL COMMITMENTS | | | |
|---|---|---|---|---|
| Item Number | Commitment | UFSAR Supplement Section/ LRA Section | Applicant's Implementation Schedule | Source |
| 1 | Enhance the Above Ground Steel Tanks Program for IP2 and IP3 to perform thickness measurements of the bottom surfaces of the condensate storage tanks, city water tank, and fire water tanks once during the first ten years of the period of extended operation.<br><br>Enhance the Above Ground Steel Tanks Program for IP2 and IP3 to require trending of thickness measurements when material loss is detected. | A.2.1.1<br>A.3.1.1<br>B.1.1 | IP2: September 28, 2013<br><br>IP3: December 12, 2015 | NL-07-039 |
| 2 | Enhance the Bolting Integrity Program for IP2 and IP3 to clarify that actual yield strength is used in selecting materials for low susceptibility to SCC and clarify the prohibition on use of lubricants containing MoS$_2$ for bolting.<br><br>The Bolting Integrity Program manages loss of preload and loss of material for all external bolting. | A.2.1.2<br>A.3.1.2<br>B.1.2<br><br>Audit Items 201, 241, 270 | IP2: September 28, 2013<br><br>IP3: December 12, 2015 | NL-07-039<br>NL-07-153 |

A-1

APPENDIX A: INDIAN POINT NUCLEAR GENERATING UNIT NOS. 2 AND 3 LICENSE RENEWAL COMMITMENTS

| Item Number | Commitment | UFSAR Supplement Section/ LRA Section | Applicant's Implementation Schedule | Source |
|---|---|---|---|---|
| 3 | Implement the Buried Piping and Tanks Inspection Program for IP2 and IP3 as described in LRA Section B.1.6.<br><br>This new program will be implemented consistent with the corresponding program described in NUREG-1801, Section XI.M34, Buried Piping and Tanks Inspection.<br><br>Include in the Buried Piping and Tanks Inspection Program described in LRA Section B.1.6 a risk assessment of in-scope buried piping and tanks that includes consideration of the impacts of buried piping or tank leakage and of conditions affecting the risk for corrosion. Classify pipe segments and tanks as having a high, medium or low impact of leakage based on the safety class, the hazard posed by fluid contained in the piping and the impact of leakage on reliable plant operation. Determine corrosion risk through consideration of piping or tank material, soil resistivity, drainage, the presence of cathodic protection and the type of coating. Establish inspection priority and frequency for periodic inspections of the in-scope piping and tanks based on the results of the risk assessment. Perform inspections using inspection techniques with demonstrated effectiveness. | A.2.1.5<br>A.3.1.5<br>B.1.6<br>Audit Item 173 | IP2: September 28, 2013<br>IP3: December 12, 2015 | NL-07-039<br>NL-07-153<br>NL-09-106<br>NL-09-111<br>NL-11-101 |
| 4 | Enhance the Diesel Fuel Monitoring Program to include cleaning and inspection of the IP2 GT-1 gas turbine fuel oil storage tanks, IP2 and IP3 EDG fuel oil day tanks, IP2 SBO/Appendix R diesel generator fuel oil day tank, and IP3 Appendix R fuel oil storage tank and day tank once every ten years. | A.2.1.8<br>A.3.1.8<br>B.1.9<br>Audit Items 128, 129, 132, 491, 492, 510 | IP2: September 28, 2013<br>IP3: December 12, 2015 | NL-07-039<br>NL-07-153<br>NL-08-057 |

## APPENDIX A: INDIAN POINT NUCLEAR GENERATING UNIT NOS. 2 AND 3 LICENSE RENEWAL COMMITMENTS

| Item Number | Commitment | UFSAR Supplement Section/ LRA Section | Applicant's Implementation Schedule | Source |
|---|---|---|---|---|
| 4 (continued) | Enhance the Diesel Fuel Monitoring Program to include quarterly sampling and analysis of the IP2 SBO/Appendix R diesel generator fuel oil day tank, IP2 security diesel fuel oil storage tank, and IP3 Appendix R fuel oil storage tank. Particulates, water and sediment checks will be performed on the samples. Filterable solids acceptance criterion will be less than or equal to 10mg/l. Water and sediment acceptance criterion will be less than or equal to 0.05%. | | | |
| | Enhance the Diesel Fuel Monitoring Program to include thickness measurement of the bottom of the following tanks once every ten years. IP2: EDG fuel oil storage tanks, EDG fuel oil day tanks, SBO/Appendix R diesel generator fuel oil day tank, GT-1 gas turbine fuel oil storage tanks, and diesel fire pump fuel oil storage tank; IP3: EDG fuel oil day tanks, EDG fuel oil storage tanks, Appendix R fuel oil storage tank, and diesel fire pump fuel oil storage tank. | | | |
| | Enhance the Diesel Fuel Monitoring Program to change the analysis for water and particulates to a quarterly frequency for the following tanks. IP2: GT-1 gas turbine fuel oil storage tanks and diesel fire pump fuel oil storage tank; IP3: Appendix R fuel oil day tank and diesel fire pump fuel oil storage tank. | | | |
| | Enhance the Diesel Fuel Monitoring Program to specify acceptance criteria for thickness measurements of the fuel oil storage tanks within the scope of the program. | | | |
| | Enhance the Diesel Fuel Monitoring Program to direct samples be taken and include direction to remove water when detected. | | | |
| | Revise applicable procedures to direct sampling of the onsite | | | |

| Item Number | Commitment | UFSAR Supplement Section/ LRA Section | Applicant's Implementation Schedule | Source |
|---|---|---|---|---|
| 4 (continued) | portable fuel oil contents prior to transferring the contents to the storage tanks. Enhance the Diesel Fuel Monitoring Program to direct the addition of chemicals including biocide when the presence of biological activity is confirmed. | | | |
| 5 | Enhance the External Surfaces Monitoring Program for IP2 and IP3 to include periodic inspections of systems in scope and subject to aging management review for license renewal in accordance with 10 CFR 54.4(a)(1) and (a)(3). Inspections shall include areas surrounding the subject systems to identify hazards to those systems. Inspections of nearby systems that could impact the subject systems will include SSCs that are in scope and subject to aging management review for license renewal in accordance with 10 CFR 54.4(a)(2). | A.2.1.10 A.3.1.10 B.1.11 | IP2: September 28, 2013 IP3: December 12, 2015 | NL-07-039 |
| 6 | Enhance the Fatigue Monitoring Program for IP2 to monitor steady state cycles and feedwater cycles or perform an evaluation to determine monitoring is not required. Review the number of allowed events and resolve discrepancies between reference documents and monitoring procedures. Enhance the Fatigue Monitoring Program for IP3 to include all the transients identified. Assure all fatigue analysis transients are included with the lowest limiting numbers. Update the number of design transients accumulated to date. | A.2.1.11 A.3.1.11 B.1.12 Audit Item 164 | IP2: September 28, 2013 IP3: December 12, 2015 | NL-07-039 NL-07-153 |

## APPENDIX A: INDIAN POINT NUCLEAR GENERATING UNIT NOS. 2 AND 3 LICENSE RENEWAL COMMITMENTS

| Item Number | Commitment | UFSAR Supplement Section/ LRA Section | Applicant's Implementation Schedule | Source |
|---|---|---|---|---|
| 7 | Enhance the Fire Protection Program to inspect external surfaces of the IP3 RCP oil collection systems for loss of material each refueling cycle.<br><br>Enhance the Fire Protection Program to explicitly state that the IP2 and IP3 diesel fire pump engine sub-systems (including the fuel supply line) shall be observed while the pump is running. Acceptance criteria will be revised to verify that the diesel engine does not exhibit signs of degradation while running; such as fuel oil, lube oil, coolant, or exhaust gas leakage.<br><br>Enhance the Fire Protection Program to specify that the IP2 and IP3 diesel fire pump engine carbon steel exhaust components are inspected for evidence of corrosion and cracking at least once each operating cycle.<br><br>Enhance the Fire Protection Program for IP3 to visually inspect the cable spreading room, 480V switchgear room, and EDG room $CO_2$ fire suppression system for signs of degradation, such as corrosion and mechanical damage at least once every six months. | A.2.1.12<br><br>A.3.1.12<br><br>B.1.13 | IP2: September 28, 2013<br><br>IP3: December 12, 2015 | NL-07-039 |

## APPENDIX A: INDIAN POINT NUCLEAR GENERATING UNIT NOS. 2 AND 3 LICENSE RENEWAL COMMITMENTS

| Item Number | Commitment | UFSAR Supplement Section/ LRA Section | Applicant's Implementation Schedule | Source |
|---|---|---|---|---|
| 8 | Enhance the Fire Water Program to include inspection of IP2 and IP3 hose reels for evidence of corrosion. Acceptance criteria will be revised to verify no unacceptable signs of degradation.<br><br>Enhance the Fire Water Program to replace all or test a sample of IP2 and IP3 sprinkler heads required for 10 CFR 50.48 using guidance of NFPA 25 (2002 edition), Section 5.3.1.1.1 before the end of the 50-year sprinkler head service life and at 10-year intervals thereafter during the extended period of operation to ensure that signs of degradation, such as corrosion, are detected in a timely manner.<br><br>Enhance the Fire Water Program to perform wall thickness evaluations of IP2 and IP3 fire protection piping on system components using non-intrusive techniques (e.g., volumetric testing) to identify evidence of loss of material due to corrosion. These inspections will be performed before the end of the current operating term and at intervals thereafter during the period of extended operation. Results of the initial evaluations will be used to determine the appropriate inspection interval to ensure aging effects are identified prior to loss of intended function.<br><br>Enhance the Fire Water Program to inspect the internal surface of the IP3 foam based fire suppression tanks. Acceptance criteria will be enhanced to verify no significant corrosion. | A.2.1.13<br>A.3.1.13<br>B.1.14<br><br>Audit Items 105, 106 | IP2: September 28, 2013<br><br>IP3: December 12, 2015 | NL-07-039<br>NL-07-153<br>NL-08-014 |

A-6

| Item Number | Commitment | UFSAR Supplement Section/ LRA Section | Applicant's Implementation Schedule | Source |
|---|---|---|---|---|
| 9 | Enhance the Flux Thimble Tube Inspection Program for IP2 and IP3 to implement comparisons to wear rates identified in WCAP-12866. Include provisions to compare data to the previous performances and perform evaluations regarding change to test frequency and scope. Enhance the Flux Thimble Tube Inspection Program for IP2 and IP3 to specify the acceptance criteria as outlined in WCAP-12866 or other plant-specific values based on evaluation of previous test results. Enhance the Flux Thimble Tube Inspection Program for IP2 and IP3 to direct evaluation and performance of corrective actions based on tubes that exceed or are projected to exceed the acceptance criteria. Also stipulate that flux thimble tubes that cannot be inspected over the tube length and cannot be shown by analysis to be satisfactory for continued service, must be removed from service to ensure the integrity of the reactor coolant system pressure boundary. | A.2.1.15 A.3.1.15 B.1.16 | IP2: September 28, 2013 IP3: December 12, 2015 | NL-07-039 |
| 10 | Enhance the Heat Exchanger Monitoring Program for IP2 and IP3 to include the following heat exchangers in the scope of the program.<br>• Safety injection pump lube oil heat exchangers<br>• RHR heat exchangers<br>• RHR pump seal coolers<br>• Non-regenerative heat exchangers | A.2.1.16 A.3.1.16 B.1.17 Audit Item 52 | IP2: September 28, 2013 IP3: December 12, 2015 | NL-07-039 NL-07-153 NL-09-018 |

| Item Number | Commitment | UFSAR Supplement Section/ LRA Section | Applicant's Implementation Schedule | Source |
|---|---|---|---|---|
| 10 (continued) | • Charging pump seal water heat exchangers<br><br>• Charging pump fluid drive coolers<br><br>• Charging pump crankcase oil coolers<br><br>• Spent fuel pit heat exchangers<br><br>• Secondary system steam generator sample coolers<br><br>• Waste gas compressor heat exchangers<br><br>• SBO/Appendix R diesel jacket water heat exchanger (IP2 only)<br><br>Enhance the Heat Exchanger Monitoring Program for IP2 and IP3 to perform visual inspection on heat exchangers where non-destructive examination, such as eddy current inspection, is not possible due to heat exchanger design limitations.<br><br>Enhance the Heat Exchanger Monitoring Program for IP2 and IP3 to include consideration of material-environment combinations when determining sample population of heat exchangers.<br><br>Enhance the Heat Exchanger Monitoring Program for IP2 and IP3 to establish minimum tube wall thickness for the new heat exchangers identified in the scope of the program. Establish acceptance criteria for heat exchangers visually inspected to include no indication of tube erosion, vibration wear, corrosion, pitting, fouling, or scaling. | | | |

**APPENDIX A: INDIAN POINT NUCLEAR GENERATING UNIT NOS. 2 AND 3 LICENSE RENEWAL COMMITMENTS**

| Item Number | Commitment | UFSAR Supplement Section/ LRA Section | Applicant's Implementation Schedule | Source |
|---|---|---|---|---|
| 11 | Deleted | Not applicable | Not applicable | NL-09-056<br>NL-11-101 |
| 12 | Enhance the Masonry Wall Program for IP2 and IP3 to specify that the IP1 intake structure is included in the program. | A.2.1.18<br>A.3.1.18<br>B.1.19 | IP2: September 28, 2013<br>IP3: December 12, 2015 | NL-07-039 |

| Item Number | Commitment | UFSAR Supplement Section/ LRA Section | Applicant's Implementation Schedule | Source |
|---|---|---|---|---|
| 13 | Enhance the Metal-Enclosed Bus Inspection Program to add IP2 480V bus associated with substation A to the scope of the bus inspected. | A.2.1.19<br>A.3.1.19<br>B.1.20<br>Audit Item 124, 133, 519 | IP2: September 28, 2013<br>IP3: December 12, 2015 | NL-07-039<br>NL-07-153<br>NL-08-057 |
| | Enhance the Metal-Enclosed Bus Inspection Program for IP2 and IP3 to visually inspect the external surface of MEB enclosure assemblies for loss of material at least once every 10 years. The first inspection will occur prior to the period of extended operation and the acceptance criterion will be no significant loss of material. | | | |
| | Enhance the Metal-Enclosed Bus Inspection Program to add acceptance criteria for MEB internal visual inspections to include the absence of indications of dust accumulation on the bus bar, on the insulators, and in the duct, in addition to the absence of indications of moisture intrusion into the duct. | | | |
| | Enhance the Metal-Enclosed Bus Inspection Program for IP2 and IP3 to inspect bolted connections at least once every five years if performed visually or at least once every ten years using quantitative measurements such as thermography or contact resistance measurements. The first inspection will occur prior to the period of extended operation. | | | |
| | The plant will process a change to applicable site procedure to remove the reference to "re-torquing" connections for phase bus maintenance and bolted connection maintenance. | | | |

# APPENDIX A: INDIAN POINT NUCLEAR GENERATING UNIT NOS. 2 AND 3 LICENSE RENEWAL COMMITMENTS

| Item Number | Commitment | UFSAR Supplement Section/ LRA Section | Applicant's Implementation Schedule | Source |
|---|---|---|---|---|
| 14 | Implement the Non-EQ Bolted Cable Connections Program for IP2 and IP3 as described in LRA Section B.1.22. | A.2.1.21<br>A.3.1.21<br>B.1.22 | IP2: September 28, 2013<br>IP3: December 12, 2015 | NL-07-039 |
| 15 | Implement the Non-EQ Inaccessible Medium-Voltage Cable Program for IP2 and IP3 as described in LRA Section B.1.23.<br><br>This new program will be implemented consistent with the corresponding described in NUREG-1801, Section XI.E3, Inaccessible Medium-Voltage Cables Not Subject to 10 CFR 50.49 Environmental Qualification Requirements. | A.2.1.22<br>A.3.1.22<br>B.1.23<br>Audit Item 173 | IP2: September 28, 2013<br>IP3: December 12, 2015 | NL-07-039<br>NL-07-153<br>NL-11-032<br>NL-11-096<br>NL-11-101 |
| 16 | Implement the Non-EQ Instrumentation Circuits Test Review Program for IP2 and IP3 as described in LRA Section B.1.24.<br><br>This new program will be implemented consistent with the corresponding described in NUREG-1801, Section XI.E2, Electrical Cables and Connections Not Subject to 10 CFR 50.49 Environmental Qualification Requirements Used in Instrumentation Circuits. | A.2.1.23<br>A.3.1.23<br>B.1.24<br>Audit Item 173 | IP2: September 28, 2013<br>IP3: December 12, 2015 | NL-07-039<br>NL-07-153 |
| 17 | Implement the Non-EQ Insulated Cables and Connections Program for IP2 and IP3 as described in LRA Section B.1.25.<br><br>This new program will be implemented consistent with the corresponding described in NUREG-1801, Section XI.E1, Electrical Cables and Connections Not Subject to 10 CFR 50.49 Environmental Qualification Requirements. | A.2.1.24<br>A.3.1.24<br>B.1.25<br>Audit Item 173 | IP2: September 28, 2013<br>IP3: December 12, 2015 | NL-07-039<br>NL-07-153 |

| Item Number | Commitment | UFSAR Supplement Section/ LRA Section | Applicant's Implementation Schedule | Source |
|---|---|---|---|---|
| 18 | Enhance the Oil Analysis Program for IP2 to sample and analyze lubricating oil used in the SBO/Appendix R diesel generator consistent with the oil analysis for other site diesel generators.<br><br>Enhance the Oil Analysis Program for IP2 and IP3 to sample and analyze generator seal oil and turbine hydraulic control oil.<br><br>Enhance the Oil Analysis Program for IP2 and IP3 to formalize preliminary oil screening for water and particulates and laboratory analyses including defined acceptance criteria for all components included in the scope of this program. The program will specify corrective actions in the event acceptance criteria are not met.<br><br>Enhance the Oil Analysis Program for IP2 and IP3 to formalize trending of preliminary oil screening results as well as data provided from independent laboratories. | A.2.1.25<br>A.3.1.25<br>B.1.26 | IP2: September 28, 2013<br><br>IP3: December 12, 2015 | NL-07-039<br>NL-11-101 |
| 19 | Implement the One-Time Inspection Program for IP2 and IP3 as described in LRA Section B.1.27.<br><br>This new program will be implemented consistent with the corresponding program described in NUREG-1801, Section XI.M32, One-Time Inspection. | A.2.1.26<br>A.3.1.26<br>B.1.27<br>Audit Item 173 | IP2: September 28, 2013<br><br>IP3: December 12, 2015 | NL-07-039<br>NL-07-153 |

# APPENDIX A: INDIAN POINT NUCLEAR GENERATING UNIT NOS. 2 AND 3 LICENSE RENEWAL COMMITMENTS

| Item Number | Commitment | UFSAR Supplement Section/ LRA Section | Applicant's Implementation Schedule | Source |
|---|---|---|---|---|
| 20 | Implement the One-Time Inspection – Small Bore Piping Program for IP2 and IP3 as described in LRA Section B.1.28.<br><br>This new program will be implemented consistent with the corresponding program described in NUREG-1801, Section XI.M35, One-Time Inspection of ASME Code Class I Small-Bore Piping. | A.2.1.27<br>A.3.1.27<br>B.1.28<br>Audit Item 173 | IP2: September 28, 2013<br><br>IP3: December 12, 2015 | NL-07-039<br>NL-07-153 |
| 21 | Enhance the Periodic Surveillance and Preventive Maintenance Program for IP2 and IP3 as necessary to assure that the effects of aging will be managed such that applicable components will continue to perform their intended functions consistent with the current licensing basis through the period of extended operation. | A.2.1.28<br>A.3.1.28<br>B.1.29 | IP2: September 28, 2013<br><br>IP3: December 12, 2015 | NL-07-039 |
| 22 | Enhance the Reactor Vessel Surveillance Program for IP2 and IP3 revising the specimen capsule withdrawal schedules to draw and test a standby capsule to cover the peak reactor vessel fluence expected through the end of the period of extended operation.<br><br>Enhance the Reactor Vessel Surveillance Program for IP2 and IP3 to require that tested and untested specimens from all capsules pulled from the reactor vessel are maintained in storage. | A.2.1.31<br>A.3.1.31<br>B.1.32 | IP2: September 28, 2013<br><br>IP3: December 12, 2015 | NL-07-039 |
| 23 | Implement the Selective Leaching Program for IP2 and IP3 as described in LRA Section B.1.33.<br><br>This new program will be implemented consistent with the corresponding program described in NUREG-1801, Section XI.M33, Selective Leaching of Materials. | A.2.1.32<br>A.3.1.32<br>B.1.33<br>Audit Item 173 | IP2: September 28, 2013<br><br>IP3: December 12, 2015 | NL-07-039<br>NL-07-153 |

| Item Number | Commitment | UFSAR Supplement Section/ LRA Section | Applicant's Implementation Schedule | Source |
|---|---|---|---|---|
| 24 | Enhance the Steam Generator Integrity Program for IP2 and IP3 to require that the results of the condition monitoring assessment are compared to the operational assessment performed for the prior operating cycle with differences evaluated. | A.2.1.34<br>A.3.1.34<br>B.1.35 | IP2: September 28, 2013<br>IP3: December 12, 2015 | NL-07-039 |
| 25 | Enhance the Structures Monitoring Program to explicitly specify that the following structures are included in the program.<br>• Appendix R diesel generator foundation (IP3)<br>• Appendix R diesel generator fuel oil tank vault (IP3)<br>• Appendix R diesel generator switchgear and enclosure (IP3)<br>• city water storage tank foundation<br>• condensate storage tanks foundation (IP3)<br>• containment access facility and annex (IP3)<br>• discharge canal (IP2/3)<br>• emergency lighting poles and foundations (IP2/3)<br>• fire pumphouse (IP2)<br>• fire protection pumphouse (IP3)<br>• fire water storage tank foundations (IP2/3)<br>• gas turbine 1 fuel storage tank foundation<br>• maintenance and outage building-elevated passageway (IP2)<br>• new station security building (IP2)<br>• nuclear service building (IP1)<br>• primary water storage tank foundation (IP3)<br>• refueling water storage tank foundation (IP3)<br>• security access and office building (IP3)<br>• service water pipe chase (IP2/3) | A.2.1.35<br>A.3.1.35<br>B.1.36<br>Audit Items 86, 87, 88, 417, 358, 360 | IP2: September 28, 2013<br>IP3: December 12, 2015 | NL-07-039<br>NL-07-153<br>NL-08-057<br>NL-08-127<br>NL-11-032<br>NL-11-101 |

# APPENDIX A: INDIAN POINT NUCLEAR GENERATING UNIT NOS. 2 AND 3 LICENSE RENEWAL COMMITMENTS

| Item Number | Commitment | UFSAR Supplement Section/ LRA Section | Applicant's Implementation Schedule | Source |
|---|---|---|---|---|
| 25 (continued) | • service water valve pit (IP3)<br>• superheater stack<br>• transformer/switchyard support structures (IP2)<br>• waste holdup tank pits (IP2/3)<br><br>Enhance the Structures Monitoring Program for IP2 and IP3 to clarify that in addition to structural steel and concrete, the following commodities (including their anchorages) are inspected for each structure as applicable.<br><br>• cable trays and supports<br>• concrete portion of reactor vessel supports<br>• conduits and supports<br>• cranes, rails and girders<br>• equipment pads and foundations<br>• fire proofing (pyrocrete)<br>• HVAC duct supports<br>• jib cranes<br>• manholes and duct banks<br>• manways, hatches and hatch covers<br>• monorails<br>• new fuel storage racks<br>• sumps, sump screens, strainers and flow barriers<br><br>Enhance the Structures Monitoring Program for IP2 and IP3 to inspect inaccessible concrete areas that are exposed by excavation for any reason. IP2 and IP3 will also inspect inaccessible concrete areas in environments where observed conditions in accessible areas exposed to the same environment | | | |

| Item Number | Commitment | UFSAR Supplement Section/ LRA Section | Applicant's Implementation Schedule | Source |
|---|---|---|---|---|
| 25 (continued) | indicate that significant concrete degradation is occurring. | | | |
| | Enhance the Structures Monitoring Program for IP2 and IP3 to perform inspections of elastomers (seals, gaskets, seismic joint filler, and roof elastomers) to identify cracking and change in material properties and for inspection of aluminum vents and louvers to identify loss of material. | | | |
| | Enhance the Structures Monitoring Program for IP2 and IP3 to perform an engineering evaluation of groundwater samples to assess aggressiveness of groundwater to concrete on a periodic basis (at least once every five years). IPEC will obtain samples from at least 5 wells that are representative of the ground water surrounding below-grade site structures and perform an engineering evaluation of the results from those samples for sulfates, pH and chlorides. Additionally, to assess potential indications of spent fuel pool leakage, IPEC will sample for tritium in groundwater wells in close proximity to the IP2 spent fuel pool at least once every 3 months. | | | |
| | Enhance the Structures Monitoring Program for IP2 and IP3 to perform inspection of normally submerged concrete portions of the intake structures at least once every 5 years. Inspect the baffling/grating partition and support platform of the IP3 intake structure at least once every 5 years. | | | |
| | Enhance the Structures Monitoring Program for IP2 and IP3 to perform inspection of the degraded areas of the water control structure once per 3 years rather than the normal frequency of once per 5 years during the PEO. | | | |

| Item Number | Commitment | UFSAR Supplement Section/ LRA Section | Applicant's Implementation Schedule | Source |
|---|---|---|---|---|
| 25 (continued) | Enhance the Structures Monitoring Program to include more detailed quantitative acceptance criteria for inspections of concrete structures in accordance with ACI 349.3R, "Evaluation of Existing Nuclear Safety-Related Concrete Structures" prior to the period of extended operation. | | | |
| 26 | Implement the Thermal Aging Embrittlement of Cast Austenitic Stainless Steel (CASS) Program for IP2 and IP3 as described in LRA Section B.1.37. <br><br> This new program will be implemented consistent with the corresponding program described in NUREG-1801, Section XI.M12, Thermal Aging Embrittlement of Cast Austenitic Stainless Steel (CASS) Program. | A.2.1.36 <br> A.3.1.36 <br> B.1.37 <br> Audit Item 173 | IP2: September 28, 2013 <br> IP3: December 12, 2015 | NL-07-039 <br> NL-07-153 |
| 27 | Implement the Thermal Aging and Neutron Irradiation Embrittlement of Cast Austenitic Stainless Steel (CASS) Program for IP2 and IP3 as described in LRA Section B.1.38. <br><br> This new program will be implemented consistent with the corresponding program described in NUREG-1801, Section XI.M13, Thermal Aging and Neutron Embrittlement of Cast Austenitic Stainless Steel (CASS). | A.2.1.37 <br> A.3.1.37 <br> B.1.38 <br> Audit Item 173 | IP2: September 28, 2013 <br> IP3: December 12, 2015 | NL-07-039 <br> NL-07-153 |

| Item Number | Commitment | UFSAR Supplement Section/ LRA Section | Applicant's Implementation Schedule | Source |
|---|---|---|---|---|
| 28 | Enhance the Water Chemistry Control – Closed Cooling Water Program to maintain water chemistry of the IP2 SBO/Appendix R diesel generator cooling system per EPRI guidelines.<br><br>Enhance the Water Chemistry Control – Closed Cooling Water Program to maintain the IP2 and IP3 security generator and fire protection diesel cooling water pH and glycol within limits specified by EPRI guidelines. | A.2.1.39<br>A.3.1.39<br>B.1.40<br>Audit Item 509 | IP2: September 28, 2013<br>IP3: December 12, 2015 | NL-07-039<br>NL-08-057 |
| 29 | Enhance the Water Chemistry Control – Primary and Secondary Program for IP2 to test sulfates monthly in the RWST with a limit of <150 ppb. | A.2.1.40<br>B.1.41 | IP2: September 28, 2013 | NL-07-039 |
| 30 | For aging management of the reactor vessel internals, IPEC will (1) participate in the industry programs for investigating and managing aging effects on reactor internals; (2) evaluate and implement the results of the industry programs as applicable to the reactor internals; and (3) upon completion of these programs, but not less than 24 months before entering the period of extended operation, submit an inspection plan for reactor internals to the NRC for review and approval. | A.2.1.41<br>A.3.1.41<br>B.1.41 | IP2: September 28, 2011<br>IP3: December 12, 2013 | NL-07-039 |
| 31 | Additional P-T curves will be submitted as required per 10 CFR 50, Appendix G prior to the period of extended operation as part of the Reactor Vessel Surveillance Program. | A.2.2.1.2<br>A.3.2.1.2<br>4.2.3 | IP2: September 28, 2013<br>IP3: December 12, 2015 | NL-07-039 |

A-18

| Item Number | Commitment | UFSAR Supplement Section/ LRA Section | Applicant's Implementation Schedule | Source |
|---|---|---|---|---|
| 32 | As required by 10 CFR 50.61(b)(4), IP3 will submit a plant-specific safety analysis for plate B2803-3 to the NRC three years prior to reaching the RT$_{PTS}$ screening criterion. Alternatively, the site may choose to implement the revised PTS rule when approved. | A.3.2.1.4 4.2.5 | IP3: December 12, 2015 | NL-07-039 NL-08-127 |
| 33 | At least 2 years prior to entering the period of extended operation, for the locations identified in LRA Table 4.3-13 (IP2) and LRA Table 4.3-14 (IP3), under the Fatigue Monitoring Program, IP2 and IP3 will implement one or more of the following:

(1) Consistent with the Fatigue Monitoring Program, Detection of Aging Effects, update the fatigue usage calculations using refined fatigue analyses to determine valid CUFs less than 1.0 when accounting for the effects of reactor water environment. This includes applying the appropriate Fen factors to valid CUFs determined in accordance with one of the following:

1. For locations in LRA Table 4.3-13 (IP2) and LRA Table 4.3-14 (IP3), with existing fatigue analysis valid for the period of extended operation, use the existing CUF.

2. Additional plant-specific locations with a valid CUF may be evaluated. In particular, the pressurizer lower shell will be reviewed to ensure the surge nozzle remains the limiting component.

3. Representative CUF values from other plants, adjusted to or enveloping the IPEC plant specific external loads may be used if demonstrated applicable to IPEC. | A.2.2.3.3 A.3.2.2.3 4.3.3 Audit Item 146 | IP2: September 28, 2011 IP3: December 12, 2013 Complete | NL-07-039 NL-07-153 NL-08-021 NL-10-082 |

APPENDIX A: INDIAN POINT NUCLEAR GENERATING UNIT NOS. 2 AND 3 LICENSE RENEWAL COMMITMENTS

| Item Number | Commitment | UFSAR Supplement Section/ LRA Section | Applicant's Implementation Schedule | Source |
|---|---|---|---|---|
| 33 (continued) | 4. An analysis using an NRC-approved version of the ASME code or NRC-approved alternative (e.g., NRC-approved code case) may be performed to determine a valid CUF.<br><br>(2) Consistent with the Fatigue Monitoring Program, Corrective Actions, repair or replace the affected locations before exceeding a CUF of 1.0. | | | |
| 34 | IP2 SBO/Appendix R diesel generator will be installed and operational by April 30, 2008. This committed change to the facility meets the requirements of 10 CFR 50.59(c)(1) and, therefore, a license amendment pursuant to 10 CFR 50.90 is not required. | 2.1.1.3.5 | April 30, 2008<br><br>Complete | NL-07-078<br>NL-08-074<br>NL-11-101 |
| 35 | Perform a one-time inspection of representative sample area of IP2 containment liner affected by the 1973 event behind the insulation, prior to entering the period of extended operation, to assure liner degradation is not occurring in this area.<br><br>Perform a one-time inspection of representative sample area of the IP3 containment steel liner at the juncture with the concrete floor slab, prior to entering the period of extended operation, to assure liner degradation is not occurring in this area.<br><br>Any degradation will be evaluated for updating of the containment liner analyses as needed. | Audit Item 27 | IP2: September 28, 2013<br><br>IP3: December 12, 2015 | NL-08-127<br>NL-09-018<br>NL-11-101 |

A-20

| Item Number | Commitment | UFSAR Supplement Section/ LRA Section | Applicant's Implementation Schedule | Source |
|---|---|---|---|---|
| 36 | Perform a one-time inspection and evaluation of a sample of potentially affected IP2 refueling cavity concrete prior to the period of extended operation. The sample will be obtained by core boring the refueling cavity wall in an area that is susceptible to exposure to borated water leakage. The inspection will include an assessment of embedded reinforcing steel.<br><br>Additional core bore samples will be taken, if the leakage is not stopped, prior to the end of the first ten years of the period of extended operation.<br><br>A sample of leakage fluid will be analyzed to determine the composition of the fluid. If additional core samples are taken prior to the end of the first ten years of the period of extended operation, a sample of leakage fluid will be analyzed. | Audit Item 359 | IP2: September 28, 2013 | NL-08-127<br>NL-09-056<br>NL-09-079<br>NL-11-101 |
| 37 | Enhance the Containment Inservice Inspection (CII-IWL) Program to include inspections of the containment using enhanced characterization of degradation (i.e., quantifying the dimensions of noted indications through the use of optical aids) during the period of extended operation. The enhancement includes obtaining critical dimensional data of degradation where possible through direct measurement or the use of scaling technologies for photographs, and the use of consistent vantage points for visual inspections. | Audit Item 361 | IP2: September 28, 2013<br><br>IP3: December 12, 2015 | NL-08-127 |

APPENDIX A: INDIAN POINT NUCLEAR GENERATING UNIT NOS. 2 AND 3 LICENSE RENEWAL COMMITMENTS

| Item Number | Commitment | UFSAR Supplement Section/ LRA Section | Applicant's Implementation Schedule | Source |
|---|---|---|---|---|
| 38 | For Reactor Vessel Fluence, should future core loading patterns invalidate the basis for the projected values of $RT_{PTS}$ or $C_V USE$, updated calculations will be provided to the NRC. | 4.2.1 | IP2: September 28, 2013<br><br>IP3: December 12, 2015 | NL-08-143 |
| 39 | Deleted | Not applicable | Not applicable | NL-09-079 |
| 40 | Evaluate plant specific and appropriate industry operating experience and incorporate lessons learned in establishing appropriate monitoring and inspection frequencies to assess aging effects for the new aging management programs. Documentation of the operating experience evaluated for each new program will be available on site for NRC review prior to the period of extended operation. | B.1.6<br>B.1.22<br>B.1.23<br>B.1.24<br>B.1.25<br>B.1.27<br>B.1.28<br>B.1.33<br>B.1.37<br>B.1.38 | IP2: September 28, 2013<br><br>IP3: December 12, 2015 | NL-09-106 |

| Item Number | Commitment | UFSAR Supplement Section/ LRA Section | Applicant's Implementation Schedule | Source |
|---|---|---|---|---|
| 41 | IPEC will inspect steam generators for both units to assess the condition of the divider plate assembly. The examination technique used will be capable of detecting PWSCC in the steam generator divider plate assembly. The IP2 steam generator divider plate inspections will be completed within the first ten years of the period of extended operation (PEO). The IP3 steam generator divider plate inspections will be completed within the first refueling outage following the beginning of the PEO. | N/A | IP2: After the beginning of the PEO and prior to September 23, 2023<br><br>IP3: Prior to the end of the first refueling outage following the beginning of the PEO | NL-11-032<br>NL-11-074<br>NL-11-090<br>NL-11-101 |
| 42 | IPEC will develop a plan for each unit to address the potential for cracking of the primary to secondary pressure boundary due to PWSCC of tube-to-tubesheet welds using one of the following two options.<br><br>Option 1 (Analysis)<br><br>IPEC will perform an analytical evaluation of the steam generator tube-to-tubesheet welds in order to establish a technical basis for either determining that the tubesheet cladding and welds are not susceptible to PWSCC, or redefining the pressure boundary in which the tube-to-tubesheet weld is no longer included and, therefore, is not required for reactor coolant pressure boundary function. The redefinition of the reactor coolant pressure boundary must be approved by the NRC as a license amendment request. | N/A | Option 1:<br>IP2: Prior to March 2024<br><br>IP3: Prior to the end of the first refueling outage following the beginning of the PEO. check these against august 9 letter.<br><br>Option 2:<br>IP2: Between March 2020 and March 2024 | NL-11-032<br>NL-11-074<br>NL-11-090<br>NL-11-096 |

## APPENDIX A: INDIAN POINT NUCLEAR GENERATING UNIT NOS. 2 AND 3 LICENSE RENEWAL COMMITMENTS

| Item Number | Commitment | UFSAR Supplement Section/ LRA Section | Applicant's Implementation Schedule | Source |
|---|---|---|---|---|
| 42 (cont.) | Option 2 (Inspection)<br><br>IPEC will perform a one-time inspection of a representative number of tube-to-tubesheet welds in each steam generator to determine if PWSCC cracking is present. If weld cracking is identified:<br><br>a. The condition will be resolved through repair or engineering evaluation to justify continued service, as appropriate, and<br><br>b. An ongoing monitoring program will be established to perform routine tube-to-tubesheet weld inspections for the remaining life of the steam generators. | | IP3: Prior to the end of the first refueling outages following the beginning of the PEO. | |
| 43 | IPEC will review design basis ASME Code Class 1 fatigue evaluations to determine whether the NUREG/CR-6260 locations that have been evaluated for the effects of the reactor coolant environment on fatigue usage are the limiting locations for the IP2 and IP3 configurations. If more limiting locations are identified, the most limiting location will be evaluated for the effects of the reactor coolant environment on fatigue usage.<br><br>IPEC will use the NUREG/CR-6909 methodology in the evaluation of the limiting locations consisting of nickel alloy, if any. | 4.3.3 | IP2: Prior to September 28, 2013<br><br>IP3: Prior to December 12, 2015 | NL-11-032<br><br>NL-11-101 |

A-24

## APPENDIX A: INDIAN POINT NUCLEAR GENERATING UNIT NOS. 2 AND 3 LICENSE RENEWAL COMMITMENTS

| Item Number | Commitment | UFSAR Supplement Section/ LRA Section | Applicant's Implementation Schedule | Source |
|---|---|---|---|---|
| 44 | IPEC will include written explanation and justification of any user intervention in future evaluations using the WESTEMS "Design CUF" module. | N/A | IP2: Prior to September 28, 2013 IP3: Prior to December 12, 2015 | NL-11-032 NL-11-101 |
| 45 | IPEC will not use the NB-3600 option of the WESTEMS program in future design calculations until the issues identified during the NRC review of the program have been resolved. | N/A | IP2: Prior to September 28, 2013 IP3: Prior to December 12, 2015 | NL-11-032 NL-11-101 |
| 46 | Include in the IP2 ISI Program that IPEC will perform twenty-five volumetric weld metal inspections of socket welds during each 10-year ISI interval scheduled as specified by IWB-2412 of the ASME Section XI Code during the period of extended operation. In lieu of volumetric examinations, destructive examinations may be performed, where one destructive examination may be substituted for two volumetric examinations. | N/A | IP2: Prior to September 28, 2013 | NL-11-032 NL-11-074 |

# APPENDIX B

# CHRONOLOGY

This appendix lists chronologically the licensing correspondence between the staff of the U.S. Nuclear Regulatory Commission (NRC) (the staff) and Entergy Nuclear Operations, Inc. (Entergy or the applicant). This appendix also lists other correspondence concerning the staff's review of the Indian Point Nuclear Generating Unit Nos. 2 and 3 license renewal application (LRA) (Docket Nos. 50-247 and 50-286) issued since the publication of NUREG-1930 in November 2009.

| CHRONOLOGY | |
|---|---|
| **Date** | **Subject** |
| 7/14/2010 | Letter from Entergy Nuclear Operations, Inc. to NRC, Indian Point, Units 2 & 3 - Amendment 9 to License Renewal Application, Reactor Vessel Internals Program, (ADAMS Accession No. ML102010102) |
| 7/26/2010 | Letter from Entergy Nuclear Operations, Inc. to NRC, Indian Point, Units 2 & 3, Amendment 10 to License Renewal Application, (ADAMS Accession No. ML102150347) |
| 8/9/2010 | Letter from Entergy Nuclear Operations, Inc. to NRC, Indian Point Nuclear Generating, Units 2 and 3 - License Renewal Application - Completion of Commitment #33 Regarding the Fatigue Monitoring Program, (ADAMS Accession No. ML102300504) |
| 2/10/2011 | 01/06/2011 & 01/12/2011-Summary of Telephone Conference Calls Between the NRC and Entergy Nuclear Operations Inc., Concerning Draft Requests for Additional Information (ADAMS Accession No. ML110180529) |
| 2/10/2011 | Letter from NRC to Entergy Nuclear Operations, Inc., Request for Additional Information for the Review of the Indian Point Nuclear generating Units Numbers 2 and 3, License Renewal Application, (ADAMS Accession No. ML110190809) |
| 4/19/2011 | Letter from NRC to Entergy Nuclear Operations, Inc., Safety Project Manager Change For The License Renewal Project For Indian Point Nuclear Generating Unit Numbers 2 And 3, (ADAMS Accession No. ML11090A078) |
| 3/28/2011 | Letter from Entergy Nuclear Operations, Inc. to NRC, Indian Point, Units 2 & 3, Response to Request for Additional Information on Aging Management Programs, (ADAMS Accession No. ML110960360) |
| 5/9/2011 | 01/12/11 Summary of Telephone Conference Call Between NRC and Entergy Nuclear Operations, Inc., Concerning the Final SEIS for the Proposed License Renewal of Indian Point, Units 2 and 3, (ADAMS Accession No. ML11102A006) |
| 6/15/2011 | Letter from NRC to Entergy Nuclear Operations, Inc, Request For Additional Information For The Review Of The Indian Point Nuclear Generating Unit Nos. 2 and 3, License Renewal Application, (ADAMS Accession No. ML11139A447) |
| 7/14/2011 | Letter from Nuclear Operations, Inc, to NRC, Indian Point Units 2 & 3, Response to Request for Additional Information Aging Management Programs, (ADAMS Accession No. ML11201A160) |

| CHRONOLOGY | |
|---|---|
| **Date** | **Subject** |
| 7/18/2011 | 06/09/2011-Summary of Conference Call Between the NRC and Entergy Nuclear Operations, Inc., Concerning Inspection of Small Bore Piping, Steam Generator Components, and the Non-EQ Inaccessible Medium-Voltage Cable Program, (ADAMS Accession No. ML11178A335) |
| 7/27/2011 | Letter from Nuclear Operations, Inc, to NRC, Indian Point Units 2 & 3, Clarification for Request for Additional Information Aging Management Programs, (ADAMS Accession No. ML11215A128) |
| 8/9/2011 | Letter from Nuclear Operations, Inc, to NRC, Indian Point Units 2 & 3, Clarification for Request for Additional Information Aging Management Programs, (ADAMS Accession No. ML11229A803) |
| 8/22/2011 | Letter from Nuclear Operations, Inc, to NRC, Indian Point Units 2 & 3, Clarification for Request for Additional Information Aging Management Programs ADAMS Accession No. ML11XXXXXXX) |
| 8/30/11 | 7/21/2011-Summary of Conference Call Between the NRC and Entergy Nuclear Operations, Inc., Concerning Operating Experience at Indian Point Nuclear Generating Station, Units 2 and 3, (ADAMS Accession No. ML11215A056) |
| 8/30/11 | 07/25/2011-Summary of Conference Call Between the NRC and Entergy Nuclear Operations, Inc., Concerning Steam Generator Aging Management and Metal Fatigue Analysis, (ADAMS Accession No. ML11215A088) |
| 8/30/11 | 8/3/2011-Summary of Conference Call Between the NRC and Entergy Nuclear Operations, Inc., Concerning the Non-EQ Inaccessible Medium-Voltage Cable Program, (ADAMS Accession No. ML112270145) |

www.ingramcontent.com/pod-product-compliance
Lightning Source LLC
Chambersburg PA
CBHW081849170526
45167CB00007B/2942